A. D. J. MEEUSE, D.Phil., University of Leyden, is Professor of Systematic Botany and Plant Geography, University of Amsterdam. He previously served as Scientific Officer at the Fibre Research Institute of the Organization for Applied Scientific Research in Delft, The Netherlands, and as Senior Scientific Officer, Division of Botany and National Herbarium, Pretoria, South Africa.

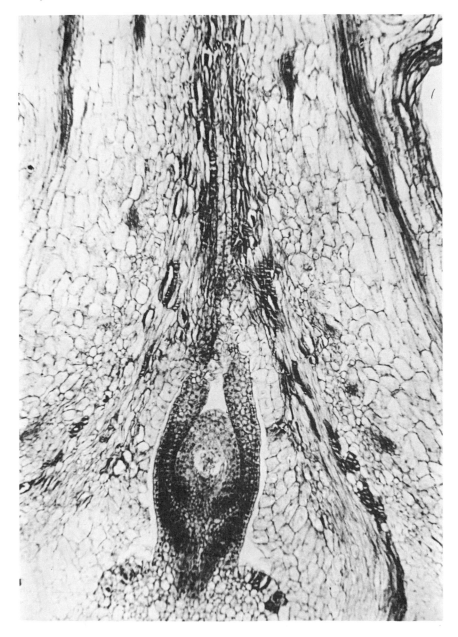

PLATE I. Part of a longitudinal section of the pistil of *Engelhardia spicata* (Juglandac.) stained with Safranin-Fast Green, showing the tubular connection between the ovule and the stigmatic area. ×125. Compare Fig. 12. (Slide prepared by J. Houthuesen, photomicrograph by W. van Leeuwen.)

FUNDAMENTALS OF PHYTOMORPHOLOGY

A. D. J. Meeuse

University of Amsterdam

THE RONALD PRESS COMPANY • NEW YORK

Copyright © 1966 by
THE RONALD PRESS COMPANY

———

All Rights Reserved

Library of Congress Catalog Card Number: 65–21813

PRINTED IN THE UNITED STATES OF AMERICA

To the Nestor of our contemporary dynamic phytomorphologists, the veritable torch bearer of the New Morphology these last thirty years, my old mentor and very good friend

Dr. Herman Johannes Lam

Emeritus Professor of Systematic Botany and Plant Geography, Late Director of the Rijksherbarium, Leyden, this book is gratefully dedicated.

Preface

More than thirty years have passed since H. H. Thomas first spoke of the 'New Morphology', which meant to convey that the study of the so-called Pteridophyta and of many fossil forms was changing the face of phytomorphology. Although the 'New Morphology' does not deserve the epithet 'brand-new' any more, the ideas expressed by Thomas have lost nothing of their actuality, and in these thirty years some phytomorphologists have started thinking along these lines. Strangely enough, the 'new' fundamentals, though perfectly sound, did not change general morphological thinking sufficiently to replace the classical concepts of the 'Old Morphology' to an appreciable extent.

It is already fifteen years since my old teacher and good friend, H. J. Lam, rather optimistically stated that the New Morphology was rapidly gaining ground and paid homage to the line of predecessors, from Bower to Zimmermann, 'who were so much in advance of their time that theirs was a *vox clamantis in deserto*'. Little could he, the veritable champion of the New Morphology, have expected the reactions of some of the leading phytomorphologists to his own voice! He was subjected to a great deal of criticism, some of which even took a form that somebody once described as 'a vitriolic attack' and was certainly unfair, to say the least. But why the abuse and why the negative stand of so many botanists who cling tenaciously and sometimes fanatically to antiquated, nay, I daresay archaic, morphological notions? The American palaeobotanist, C. A. Arnold, has criticised the attitude prevailing among a certain category of systematists, to whom a scientific theory, once it has become widely adopted, 'takes on many attributes of a creed' and to whom the framework of their thinking becomes a substitute for the preconceived working hypotheses: 'It [*i.e.*, the theory] comes to be looked upon as having emanated from some authoritative and inspired source. It is accepted as final, and anyone who would be so bold as to suggest that it be altered to conform with modern knowledge is promptly squelched with a smooth flow of well-rehearsed oratory.' As regards phytomorphologists—whom the cap fits, let him wear it. Arnold's sally certainly explains, to my mind, the fairly general dissent from novel suggestions and the treatment of the originators of such shocking ideas as nosey parkers, if not as mad bulls in Ye Olde Morphological China Shoppe.

v

Apart from this deplorable attitude, there are the conservatism, the upbringing in some particular traditional morphological 'school', and, perhaps, a certain mental laziness of many workers that prevents them from adopting the New Morphology without the greatest possible reluctance. It is true, as LAM and ZIMMERMANN have pointed out, that everybody who was brought up in the Old Morphology—a category which, incidentally, includes my mentor, LAM, and myself—at first finds it difficult to disengage himself from the tenets ingrained in his mind. However, as long as one is determined to give it a serious try (or should I say, daring enough to shake off the shackles?), it becomes easier to rid oneself of the established, so-called 'classical' morphological doctrines as one proceeds along the new road. It is certainly a much more adventurous trail to reconnoitre.

The Old Morphology, built up by a number of outstanding botanists over a period of more than a century, is a formidable edifice, and it would be foolish to declare that it would have to be completely demolished and replaced by a New Phytomorphology. Many of its fundamentals, such as the concept of homology, are essentially sound and paved the way for the typological systems of classification which are used for our standard taxonomic manuals, monographs, and regional floras. It is only when the Old Morphology obviously has become too static and too dogmatic, when it ignores new evidence, and when it hampers the progress of science that it has to be drastically emended.

It is my firm conviction that the rather unfavourable reputation of the so-called non-experimental (morphological and taxonomic) biological disciplines among workers in the so-called experimental branches is partly to be ascribed to the rigid application of theories developed before the turn of the century and adhered to essentially unchanged. That progress in some fields was lamentably retarded is manifestly demonstrated by the endeavours to solve DARWIN's 'abominable mystery'—the vain quest for the botanical 'missing link', the elusive ancestors of the Angiosperms. I believe I can state without gross exaggeration that the back of the problem of the origin and descent of the Flowering Plants might have been broken at least twenty-five years ago, if only this poser had been tackled in a different way. The ruling morphological train of thought put nearly all authorities on the wrong track, and the botanical literature was thus 'enriched' with a multitude of papers concerning this traditional moot point in phylogenetic botany—over fifty of them dealing with the subject in general and many more with certain aspects or details of the problem. Many of these are highly conjectural; others are mere repetitions or summaries; and some are honest admissions of defeat. The baffling inference, that fossil Pre-angiosperms could not be found or indicated, in turn led to various circumstantial explanations as to why such fossils were not likely to be encountered. The obvious but rather embarrassing conclusion is that one was looking for a hypothetical type of plant that has probably never existed and, what is far worse, rejected some suggested 'possibles' and 'probables' because they did not come up to the preconceived imaginary plant form one anticipated finding.

Similarly, current notions about 'primitive' and 'advanced' ('derived') characters have saddled us with scores of typological, *i.e.*, pseudo-phylogenetic, genealogies, 'derivations' and systems of classification. Again, there is a great deal of chaff and very little grain. The botanical literature is overburdened with such contributions; they fill many pages in our journals; they clutter up our libraries; but has there been an appreciable net scientific gain? It is sad to reflect that most of the publications under discussion are partly or completely obsolete, and obviously something has to be done to curtail this still-flowing stream of rather useless information.

LAM has repeatedly called attention to the fact that the Old Morphology is 'Angiosperm-centred', in the sense that morphological concepts, conclusions and generalisations were deduced from a study of the Angiosperms (which is also reflected in the terminology) and subsequently applied to all other Cormophyta. A more logical procedure would be to develop our theories on the basis of a comparative analysis of all cormophytic groups, preferably including early extinct fossil representatives, before applying our findings to the more specialised taxa such as the Angiosperms. The morphology of the Angiosperms should be the final stage of a comparative morphological analysis, rather than its starting point. It is not altogether surprising that the application of the new phytomorphological principles to the morphology of the Angiosperms, *i.e.*, to the principal subject matter of the Old Morphology, met with the greatest resistance. Even ZIMMERMANN's *Die Phylogenie der Pflanzen* (1959), so admirable a book in several respects, offers virtually nothing new in the section dealing with the Angiosperms, and TAKHTAJAN's *Evolution der Angiospermen*, in spite of its ambitious title, does not contain anything tangible either.

The New Morphology of the Angiosperms suffers from the disadvantage that it has never been comprehensively treated in textbook form. Nearly all of our leading handbooks discuss the Angiosperms in terms of the classical morphology, and the New Morphology, if mentioned at all, is only fleetingly dealt with. Our university students cannot easily become acquainted with alternative ideas which would develop a desirable critical attitude.

These essays are a first attempt to provide some of the applications of 'new' phytomorphological concepts to the Angiosperms in a succinct form. Apart from some general and fundamental considerations, only the most controversial subjects, especially the nature of the reproductive organs in the Higher Cormophyta, are more extensively treated.

The author can only claim that the New Morphology provides us with a tool for a different approach to morphological and taxonomic problems, which not only leads to some completely unexpected conclusions but also seems to yield a satisfactory answer in cases where the Old Morphology has failed.

A. D. J. MEEUSE

Amsterdam
September, 1965

Technical Introduction and Acknowledgements

This book is not written for the beginner but expects from the prospective user somewhat more basic knowledge than a mere sprinkling of morphology, anatomy, embryology, palynology, taxonomy and palaeobotany.

To avoid too many cross-references, each chapter is somewhat of the nature of a complete essay. The indulgent reader will, I hope, prefer some inevitable repetitions to too-frequent leafing backward and forward through the pages. The bibliography is restricted almost exclusively to publications cited in the text, the numerous comprehensive manuals that cover the various disciplines involved being omitted for obvious reasons. Extensive surveys of the literature on phytomorphology and phylogeny can be found in such recent works as TAKHTAJAN's *Evolution der Angiospermen* and ZIMMERMAN's *Phylogenie der Pflanzen,* and in several earlier complications, such as CONSTANCE's exhaustive summary of the phylogenetic botany of the Angiosperms and FLORIN's equally well documented corresponding treatment of the Gymnosperms published in the same volume.

Some problems dealt with rather concisely in the present book have been discussed by the author in greater detail elsewhere, and these papers should be consulted when necessary. Other subjects not touched upon have purposely been omitted simply because it is felt that they are not controversial and hence irrelevant within the present compass.

I wish to thank all those staff members of the Hugo de Vries Laboratorium who so obligingly co-operated, especially in the more tedious chores that go with making a manuscript ready for the press. The figures and diagrams other than those originally designed for this book have been adapted mainly from papers by the author. The assistance of our artist, L. VUYK, is gratefully acknowledged. The photographs were selected from a number of originals kindly put at my disposal by my colleague, Dr. C. G. G. J. VAN STEENIS, of the Flora Malesiana Foundation, and by a few other fellow-botanists.

My wife has supported me by showing much understanding and even greater patience, apart from contributing valuable advice. Her encouragement greatly facilitated the completion of a task which I had, rather presumptuously, undertaken in an unthinking moment.

A. D. J. M.

Contents

FUNDAMENTALS OF
PHYTOMORPHOLOGY

1

On Scientific Method
and Morphology

Science is a product of man's mind, the pinnacle of his inborn desire to increase his knowledge. Scientific thinking started as soon as primitive man could correlate certain phenomena and managed to find some sort of an explanation for the interrelation of his observations, which is the essence of causality. We still follow essentially the same procedure, that is, we assemble and sort out a number of observed 'data' and produce a working hypothesis (also called a 'theory') to fit the observations into a logical framework of causality. The so-called exact sciences differ from other forms of science only in a single respect, viz., in that the former concern themselves with the phenomena as they occur in the universe independently of the activities of mankind; it is a fallacy to presume that they are more 'exact' or more 'scientific' than other scientific branches, such as those of the humanities. Far too many people suffer from the illusion that there are several different scientific methods. The 'exact' scientist often thinks that only his methods are truly scientific and looks down condescendingly on what others do as a kind of pastime, perhaps pleasant, but rather useless. This same smugness is not infrequently prevalent among biologists working in the so-called experimental disciplines (physiology, biochemistry and the like); they often feel a cut above their colleagues doing morphological and taxonomic research, although as a rule their judgment is not based on a thorough knowledge of our *scientia amabilis*. It must be admitted that the methods of morphology, taxonomy, phylogeny and related subjects are rather similar to those of the humanities (there is, among other things, a more than superficial resemblance between phylogenetic taxonomy and comparative philology), because on the whole the planned experiment is lacking; this lack is the only reason why, in some circles, these fields are not considered 'exact'. Everybody has a natural aptitude or a preference for a certain kind of work, and it is only human to regard one's own vocation as of paramount importance and somebody

3

else's with a grain of contempt; but condemning without understanding is anything but 'scientific'.

Random tests among my students revealed that many, especially the younger ones, did not have much understanding of the fundamentals of scientific thinking and confused postulates, working hypotheses, observations and 'facts'. During some of our discussions students remarked: 'That is not science but philosophy!' When I asked whether philosophy was not a science, they did not have a ready answer; yet I was not at all sure that I had convinced them. The uncomfortable feeling that they might not approach their own research or teaching with a sufficiently open mind compels me to devote some space to this topic.

The human mind works in such fashion that we cannot start from scratch to build up a mental picture but must have something tangible. We collect a number of ideas that we accept as primary and fundamental, but should not tacitly regard as 'facts', as is so often done.

These selected data are our *axioms, postulates, basic assumptions, fundamental principles,* or whatever we wish to call them, and we use them to build up a *working hypothesis* (or 'theory'). From observations we obtain a body of data (also very often erroneously regarded as 'facts'!), and these are subsequently confronted with our hypothesis; and *vice versa*. This *interpretation* of the observations must lead to a satisfactory explanation of these data by means of the theory, but as soon as we find that our data do not consistently fit into the framework of the working hypothesis, we realise that there must be something wrong with the latter and that it has to be emended or altered to fit the observed data. This adaptation of the theory should not be pursued to the limit of saving the theory at all cost by squeezing and moulding our data into the strait-jacket of a rigid working hypothesis. There comes a time when the theory, *and often also its postulates,* must be completely overhauled and, if necessary, replaced by a new one based on a different set of fundamentals. Ultimately, a stage of such perfection may be reached that the theory and the observations dovetail so satisfactorily that one gets the feeling of having attained finality and 'arrived at the truth'. However, such complacency is dangerous. In some of the exact sciences, notably in chemistry and physics, a great many of the postulates and working hypotheses have so often been confronted with new evidence that they have reached the stage of being able to survive many more such challenges unscathed; influenced by such displays of stability over considerable periods, workers in these fields readily lose their sense of discrimination, identifying the 'final' working hypothesis with 'fact' or 'truth'. The subtle but fundamental difference between ultimate hypothesis and 'fact' is generally not felt, because our methods of teaching exact sciences at the secondary schools suggest to the pupils that they are learning 'facts' and our systems of university training in the science faculties do not exactly aim at undeceiving the students in this respect.

The result is that many biologists also accept these physical and chemical 'facts' as gospel truth, simply because they are familiar and have been ingrained in their minds. It gives them that false feeling of self-esteem, of doing 'exact' scientific work, whereas unfamiliar, purely biological working hypotheses (such as, for instance, the doctrine of evolution) are suspiciously regarded as 'not proven' and hence, as untrustworthy. However, in the last decade biochemists have developed a theory of the origin of life which is based on a physicochemical working hypothesis, and thus phylogeny has come within the scope of an 'exact' experimental scientific discipline. If events take their logical course it is to be expected that, in a not too distant future, evolution will be taught and accepted as 'fact'. It is perhaps not generally known that a theory of the origin of life which is fundamentally the same as the modern biochemical one was developed more than eighty years ago, that is, at a time when plant physiology, chemistry and physics were still in their infancy, and such disciplines as biochemistry and biophysics had not even been born. Satisfying as it is to observe that this essentially biological theory has now received support from an altogether different quarter, an 'exact' and 'experimental' one into the bargain, it means only a vindication of the *biological* theory of evolution and shows that the biological working hypothesis was essentially sound. The biochemical approach is no more exact or scientific than the biological one.

On the other hand, morphologists and certain categories of systematists have made the same elementary mistake of taking the working hypothesis for 'fact', as ARNOLD (1948) has very aptly demonstrated (in a paper from which I have quoted some relevant sentences in the Preface). Several morphological doctrines have outlived the accumulation of new evidence only because the observations were interpreted to fit, or were even forcibly moulded into, the framework of the theory, whereas the theory remained virtually unaltered. This is putting the cart before the horse, and any criticism condemning this attitude is deserved, whether morphologists of the Old School like it or not. Such criticism does not so much concern those workers who clearly state their fundamentals and aims, particularly those who base their working hypothesis solely on typology, such as my countrymen BREMEKAMP (1956, 1962), the late and talented DANSER (1950) and, to a certain extent, also the late Professor PULLE of Utrecht. DANSER especially has explicitly explained that typology, if applied in its pure form, *i.e.*, without phylogenetic implications, can be a satisfactory basis for a practical system of morphological, as well as of taxonomic, classification. It remains to be seen whether this is quite true, even though one restricts the taxonomic classification to recent taxa only and the morphological one to the Angiosperms; but DANSER's approach is certainly not unscientific. However, many purely typological systems of classification have been worked out that were given the appearance of a phylogenetic basis without any tangible evidence, and it is against such unscientific methods that DANSER, BREMEKAMP and others take a firm stand. ZIMMERMANN coined one of those very descriptive words to which the German language lends itself so admirably when he referred to such pseudo-phylogenies on a typological basis as '*Alluvialphylogenien*'. The most extreme expression of the purely typological morphology I ever heard was from a morphologist, now long deceased, of the Old School who said: 'Morphology is so beautiful, because it is so complete in itself, so self-contained, and everything fits so nicely into place'. There are still many workers of this kind, who perpetrate their versions of morphology and taxonomy in this complacent way, as *l'art pour l'art*, and form a select but isolated community. This attitude may easily degenerate into an overrating of their methods and achievements, apart from so readily becoming far too rigid, too traditional and too dogmatic. This in turn leads to such deplorable scientific excesses as the worship of their doctrines as if they were the unassailable confessions of a creed and to the negation or firm rejection of any contribution that, however timidly, deviates from their own well-trodden ways of reasoning.

Another not uncommon mistake is that postulates and observations are not clearly distinguished. An observation must be independent of the postulates; there is very little point in finding agreement between certain data, which are only postulates in disguise, and the theory based on these postulates. It does not prove anything and certainly does not help to put the working hypothesis to the test.

The New Morphology must avoid all these pitfalls, and its adherents must, therefore, strive for the proper scientific approach to the subject. The postulates and theories must be clearly defined and not be given a chance of becoming conventional and obsolete.

A few final remarks about the distinction between 'experimental' and 'non-experimental' (morphological and taxonomic) disciplines, which is all the vogue nowadays. I have on a previous occasion put the rhetorical question whether a student of microscopical structures who augments his visually procured data by means of certain chemical tests on, say, cellulose, lignin, starch, etc., is just 'making observations' or 'carrying out experiments'. There is also a new and promising field of research which is called *experimental morphology*, whilst modern systematists and morphologists make good use of many data from 'experimental' disciplines such as plant physiology, genetics, biochemistry (*e.g.*, serology) and phytochemistry. The distinction is indeed vague and rather meaningless.

It is irrelevant how we procure our data, as long as the proper scientific method of interpretation of our observations is applied. The planning of an 'experiment' is essentially a prediction of a correlation deduced from the working hypothesis, but the indication of that correlation is not the monopoly of the experimentator: correlations can also be deduced from non-experimental evidence. I once came to the conclusion during monographic studies that a number of species had to be segregated and combined into a new genus. The fruits and seeds were known only from a single species; I predicted that the fruits and seeds of the other species would be found to show a number of rather singular characters, basing this deduction on the working hypothesis that the species are closely related (congeneric) and must show a correlation of taxonomic features. When eventually I procured the previously unknown fruits, they indeed proved to possess the characters I had anticipated. This strengthened my contention that these species constitute what the taxonomist calls a 'good' (or 'natural') genus. The palaeontological reconstruction of extinct fossil forms from fragmentary remains is a kind of 'prediction in retrospect'. More examples of this kind could be given, but I shall mention only one, from the humanities. On several occasions philologists have been able to reconstruct incompletely known dead languages on the basis of comparative linguistic studies which had given them such an insight into syntax and grammar that they could confidently fill in the gaps.

I hope I have convinced the reader that the methods of the 'experimental' and 'non-experimental' branches are essentially the same and that the conclusions of the 'experimental' workers are not necessarily more exact, more reliable or more 'scientific'. Physiology and biochemistry borrow heavily from chemistry and physics, apply physicochemical theories of long standing and strive at studying single processes and substances. They often work on a quantitative basis, but although numbers are heuristically rather impressive, they only represent observations.

Morphologists and taxonomists have developed their own hypotheses without borrowing from other disciplines and yet they work with much more intricate entities (cells, tissues, organs, whole organisms and species), which, as every student with a sprinkling of genetics knows, almost invariably show a great deal of genotypic and phenotypic variation. The rate of progress in phytomorphology may be slow and its results not so spectacular, but advances have been made and will continue to be made if we stick to scientific principles and if we are prepared to adopt a dynamic attitude.

2

The Old Morphology and the New

It is not my intention to discuss the Old Morphology at length in this chapter, but some of its basic principles must be indicated. However, it may not be a simple matter to state explicitly all morphological postulates, because, unlike other branches of science, many of its fundamentals were originally based on commonplace notions such as a 'leaf' and a 'flower' in the trivial meaning they had in colloquial speech. Atoms, molecules and π-mesons are not exactly household words (at least not before 1945), but it is common knowledge that a 'plant' has roots, stems, leaves, flowers, fruits and seeds. The 'plant' in question must be understood to mean 'higher plant' or 'Angiosperm', since the layman did not and very often still does not consider such things as algae, mosses, and Pteridophytes to be real 'plants'. The older botanists did not properly understand the nature of the Cryptogams and the Pteridophytes, so that these major groups were neglected or disregarded—the main reason why the Classical Morphology was and still is Angiosperm-centred. The first postulate apparently was that there are a number of categories of organs or organ systems (stem, leaf, flower, etc.). It is still a characteristic of the Old Morphology that at least some of these categories are regarded as discrete and usually mutually exclusive: hence the rigidly maintained distinction between 'axial' organs ('stems' and 'roots') and 'lateral' organs ('leaves'). It was fairly soon understood that the roots were only a special kind of 'stem'. The mutual exclusion of the two remaining categories was, among other things, supposed to be evident from the relative position of a 'leaf' as a lateral outgrowth (or derivative) of an axis and from the development of a new axis as an axillary bud in the 'axil' formed between a lateral organ and its supporting axis.

The most debatable and most debated postulate, which goes back to GOETHE's idea of 'metamorphosis', is the assignment of the fertile organs to the organ category of the 'leaves'. Thus a fertile region of a plant (read: Angiosperm), regarded as a 'flower', was seen as a monaxial system consisting of a stem bearing a number of lateral or 'appendicular' organs (leaf homologues), of which some are non-fertile (or 'sterile') and the remainder represent the male and female reproductive organs, the stamens and the 'carpels'.

I think that only one additional postulate was required to build up the Old Morphology: the concept of homology (to be discussed in detail in Chapter 4). This idea was conceived long before evolutionary theories had become common property and, originally, had no phylogenetic implications. The observers more or less intuitively felt that, in spite of the considerable variation within a certain morphological category of organs, there was some connection or relation between the organs belonging to this category, some interlinking or binding principle. This variety within the category was explained as variation on a common theme, as if some mastermind had produced (or created) a great number of possible executions from a basic pattern or a standard 'blueprint' common to them all, or even as various expressions of an idea ('idealistic' morphology). Provided one could indicate the fundamental pattern or at least grasp the basic 'idea', all actually occurring forms of one morphological category could be arranged in a more or less continuous series of increasing complexity or of increasing simplification ('reduction'), the forms at the end of such a series supposedly representing the most 'derived' (or 'advanced') forms, and those at the beginning being regarded as 'original' (or 'primitive')—these qualifications without any phylogenetic connotations. These notions are still the basis of typology in its original pure, *i.e.*, non-phylogenetic, form and are still useful as a practical tool in honest typological classifications such as advocated by DANSER. The idealistic morphology is still kept alive by TROLL and some of his pupils, who indulge in elaboration of a number of abstractions (ideas!) and have built up a morphology so mystical (verging on the mythical) that it may impress itself upon an unsuspecting observer as a form of natural philosophy, but is apparently not regarded as genuine by true philosophers.

If we now jump to the turn of the century, when the Old Morphology was brought into its present form, it is almost startling to realise that, in the course of eighty years or so, phytomorphological concepts had hardly changed at all. Even the theory of evolution, which had had such a profound effect on biological thinking, did not alter the face of morphology appreciably. Without much exaggeration, one can state that typology was still the basis of morphology and had only been given a phylogenetic disguise. A typological series of homologous organs was replaced—in name only—by a 'phylogenetic' series, hardly, if at all, transformed in other respects. Phylogeny seemed to fit so nicely into the typological framework of phytomorphology that this substitution was almost a foregone conclusion; certainly, it did not cause much of a stir. The main topic of discussion had been centred around the purely morphological 'sporo-phyll' concept and included the question of the morphological nature of the ovule. The idealistic morphology had, in correspondence with GOETHE's *Metamorphosentheorie*, postulated the homology of fertile or-gans with 'leaves'. General considerations and original research had caused serious doubt about the validity of this homology in the minds of a number of workers, such as PAYER (1857), SCHLEIDEN and SACHS. We need not go into the various arguments for and against, some of which will be discussed in another chapter, but it was undoubtedly a step backward when the 'sporophyll' theory ultimately came out the winner. It was on slender and mainly teratological evidence that ČELAKOVSKY (1874–99) defended the old concept of the *'blattbürtigen Eichen'*, but his conclusions were ultimately accepted as final by his contemporaries. To this day, they are still adhered to by protagonists of the Old School (see, *e.g.*, EAMES 1961).

One can easily find extenuating circumstances to explain why the con-servative morphological concepts came off triumphant. Typology had provided a satisfactory basis for taxonomic classification. Because it worked so well in practice, it came to be regarded as 'fact'. Palaeobotanic studies, remarkably rewarding though they had been, had not produced a clue to the origin of the Angiosperms. Even the Pteridosperms were 'recognised' only shortly after 1900! The morphology of the Pteridophytes, of the Cycads, of the Conifers, both fossil and recent, had also made good progress, but there were still considerable morphological 'gaps' between these groups and between them and the Angiosperms, so that there were no continuous series of fossils to serve as a basis for phylo-genetic lineages and true organ homologies throughout these groups. The more or less 'isolated' position of each of these major taxa did not make it easy to extend the morphological concepts and terminology, the latter 'Angiosperm-centred' of old, across all cormophytic groups. Phytomorphological terminology remained conventional and its extra-polation into other cormophytic groups necessarily somewhat forced, whilst the pteridologists were nearly as 'isolated' from the morphologists as their pet groups from the Angiosperms. Nevertheless, it was most un-fortunate that the traditional concepts about floral morphology were so universally accepted, because I cannot banish from my mind the irk-some feeling that the sporophyll concept has hampered progress not only in phytomorphology but also in taxonomy and palaeobotany. It is also strange that all criticism so suddenly ceased about 1900, although some morphological problems of long standing had not been solved, or not convincingly solved. The nature of the female coniferous cone, the oc-currence of epipetalous stamens, obdiplostemony, *'dédoublement'* (dupli-cation), centrifugal development of stamens, and the interpretative morphology of integuments, arilli, arillodes and similar organs are the most important of these 'notorious' riddles, some of which required complicated and sometimes rather far-fetched interpretations, such as a sudden 'splitting' or multiplication of floral elements, and the vanishing of whole whorls without leaving a trace.

The first fundamental 'novel' ideas of the New Morphology did not concern the fertile region. LAM (1948) has given a concise but comprehensive survey of their development, starting with BOWER's idea of 1884 about a common origin of stem and leaf, and, through the publications of H. POTONIÉ (e.g., 1912), TANSLEY and LIGNIER, leading to the first milestone, ZIMMERMANN's telome theory of 1930. A more detailed discussion, especially concerning H. POTONIÉ's ideas and publications, was given by R. POTONIÉ (1959). ZIMMERMANN cleverly combined LIGNIER's 'cauloid' theory with POTONIÉ's principle of 'overtopping' and with KIDSTON and LANG's discovery of well-preserved Devonian Psilophytes. Thus the old and rigid morphological postulate of the two independent and mutually exclusive organ categories of 'leaf' and 'stem' was completely undermined. ZIMMERMANN gradually extended and improved his theory, the final version of which was published in 1959,[*] but he adheres most emphatically to the classical sporophyll concept. In the meantime SAHNI had divided the gymnospermous groups into two categories, the Phyllospermae and the Stachyospermae, the former supposed to have leaf-borne marginal megasporangia and the latter axis-borne terminal sporangia. This idea was adopted and extended by LAM, who in 1948 divided all Cormophyta, including the Angiosperms, into phyllosporous and stachyosporous taxa. This classification cuts through traditionally uniform and supposedly homogeneous major taxonomic units ('natural' groups), separating even the male and female sexes of the same taxon. Most important was the recognition, partly based on evidence obtained by HAGERUP (1934–38), McLEAN THOMPSON (1934–37), FAGERLIND (1946, 1958) and others, that in some angiospermous taxa the ovules are axis-borne, which implies that the Flowering Plants are not monophyletic (monorheithric sensu LAM), but must have developed polyrheithrically. A hectic period of disputation ensued. LAM defended his point of view in a series of papers and concluded that he had at least the satisfaction of having provoked a great deal of inquiry, so that phytomorphology is now in a state of flux. Indeed, the classical floral morphology has repeatedly been under attack, not only by morphologists such as OZENDA (1948), NOZERAN (1955), MAEKAWA (1960), MOELIONO (1959), and MELVILLE (1900–63), but also by students of morphogenesis and ontogeny such as PLANTEFOL (1948) and BUVAT (1952, 1955).

[*] A later version (released while our work was in press) appeared as Die Telomtheorie. See Bibliography for details.

I have gone a bit further than LAM and rejected the sporophyll concept altogether. This does not imply that I accept the apparent alternative of stachyospory, because the cycadopsid ovule is essentially *cupule*-borne (MEEUSE 1964b).

The contributions from palaeobotanists in the last decades have been very important. Apart from the accumulation of new data and the recognition of several previously unknown gymnospermous groups, the most spectacular result has been FLORIN's clarification of the old problem of the coniferous cone (which was interpreted by LAM in terms of the New Morphology). However, the palaeobotanic records did not materially assist in solving those problems of floral morphology which almost coincide with the burning question of the origin of the Angiosperms. Again, I have developed certain ideas in this field and it would seem that these two interlinked problems can be successfully tackled.

When we now attempt to ascertain which are the fundamentals of the New Morphology, it appears that the main postulate is the general biological theory of phylogeny (evolution). It is assumed that the Cormophyta (or Tracheophyta) developed from non-cormophytic, thallose and unvascularised ancestors, passed through a telomic stage and subsequently developed, presumably polyphyletically, into a number of major groups (Lycophyta, Sphenophyta, ferns, Progymnosperms, etc.). In each of the major taxa the fossil records, in their stratigraphical sequence, provide us with stages in the phylogenetic development of organs and organ systems (*Merkmalsphylogenie* or semophylesis *sensu* ZIMMERMANN, 'phylogeny of single features' or organogeny *sensu* LAM). The underlying working hypothesis is simply that it is possible to recognise group or taxa phylogenies (evolutionary lines, genealogical lineages, genorheithra *sensu* LAM) which provide the true sequence of the organ phylogenies (semophyleses). This almost automatically provides the homology concept of the New Morphology: each semophylesis is essentially a series of homologous stages or phases of the organ, placed in proper sequence from the most primitive condition in the oldest taxon of the series to the more advanced forms, in the order determined by the relative age of the geologically younger taxa. Thus the organ phylogenies can be traced, and organs or associated organ systems can be related as regards their origin. Unlike typologically deduced series of homologous organs, phylogenetically reconstructed semophyleses can only be 'read' in one direction and do not, or at least not readily, lead the worker to draw up a phylogenetically speaking fictitious sequence of organs supposed to be homologous (pseudo-homologies).

The principal postulate and the ensuing working hypotheses of the New Morphology are clear and simple, but there are several practical difficulties. The fossil records are scanty and preserved partly as impressions without retained anatomical structure, partly as unconnected fragments referred to provisional 'form genera', very few forms and groups being externally and internally more or less completely known. The state of preservation also varies a great deal, so that there is much scope for interpretations and 'reconstructions'. The presence of 'gaps' between certain known forms which are presumed to belong to the same phylogenetic lineage requires additional assumptions and may result in speculations such as the suggestion of hypothetical intermediate forms. If the supposed phylogenetic relation is based on a superficial resemblance between organs of the same category whilst the other organs are disregarded, or if there is too much preconception altogether, the morphological literature is 'enriched' with another 'derivation' lacking supporting evidence and bound to be proved wanting. The Protangiosperm suggested by ARBER and PARKIN (1908) was deduced from three postulates: (1) that the Angiosperms are monophyletic, (2) that their ancestors had 'Ranalian' and Bennettitalean affinities, and (3) that these progenitors had bisexual 'strobili' with numerous 'sporophylls' which in both sexes were leaf-like (laminose) and bore marginal sporangia (or thecae and ovules, respectively; but in conventional morphological discussions the leaf-borne sporangium and the leaf-borne theca or ovule are not sharply distinguished—if the sporangium were primarily leaf-borne, whence did the ovular integuments arise?). I am willing to admit only that Protangiosperms had Bennettitalean affinities (although in a wider sense than ARBER and PARKIN supposed), and if a group of plants closely conforming to their preconceived type ever existed, which I consider doubtful, it could at best be the ancestral group of a limited number of angiospermous taxa. THOMAS (1931) thought at one time that he had at last found the elusive Pre-angiosperms in the Caytoniales and tried to relate the angiospermous 'carpel' more or less directly with the female reproductive organs of this fossil group. This is an example of a derivation centred around a possible semophylesis of a single organ or organ system, for it has since been shown that the Caytoniales are too 'primitive' and too different in several other respects to qualify as a possible ancestral group immediately preceding the Angiosperms. When GAUSSEN 'reconstructs' a semophylesis of an Angiosperm 'carpel', starting from a pteridospermous fertile frond of the *Pecopteris* type, he starts from the presumption that the Angiosperms rooted in the Medullosae. There is not the slightest palaeobotanic evidence of such a direct phylogenetic connection and indeed current opinion declines such a direct descent, so that his reconstruction with the suggested intermediate stages (in fact, a semophylesis) is completely fictitious.

A phylogenetic 'derivation' of one group from another should be based

on the correspondence of as many taxonomic, morphological, anatomical, palynological and other relevant features as can be assembled. Nobody would, for instance, seriously object to the suggested main phylogenetic history of the Coniferopsida, starting from BECK's (1960) Progymnospermopsida as the basic ancestral group and leading along several
parallel evolutionary lines to Gingkoales, Taxales, and Pinales, respectively, with the Cordaitales as an old and extinct group closely related
to but presumably running parallel with the Protopinales. There is sufficient corroborative evidence from taxonomic, morphological, palaeobotanic and other sources to make us feel confident that the data from the
various disciplines dovetail so satisfactorily that, as I have attempted to
demonstrate (MEEUSE 1963b), they can all be welded together into one
comprehensive theory of the phylogenetic, taxonomic and semophyletic
relationships not only in the Coniferopsida as a whole but also in some
of the individual main lines of descent within this class. The elucidation
of the morphological nature of the female coniferous cone is at present
one of the few textbook examples of the mutually corroborative findings
of taxonomists, palaeobotanists and phytomorphologists. FLORIN's reconstructions (1951) show clearly that such an approach to morphological problems can be highly rewarding—LAM (1954) has praised FLORIN's
work as the most spectacular advance in morphology since HOFMEISTER—
and that is why a certain measure of boldness in our suggestions is permissible. The phytomorphologist may provide a tentative semophyletic
relation which gives the palaeobotanist a lead, and *vice versa,* so that
mutually corrective studies will result and evidence from the one discipline will soon explode any exaggerated or false claims hailing from the
other (see also Chapter 6).

Considering that the fossil records are scanty and often 'far apart',
whilst the phylogenetic development of major groups often proceeds
along a number of parallel evolutionary lines with numerous ramifications, we must bear in mind that a suggested semophylesis based on a
number of scattered 'milestones' in the form of a limited number of
satisfactorily preserved fossil forms does not mean that all our records
belong to a single line of descent, but rather to several of the parallel
smaller lines within the 'over-all' phylogeny of the group under consideration. Such an assumed semophylesis is, therefore, in most cases only an
approximation of the development of characters in the group as a whole,
and we do well to realise that as a rule we have evidence only of certain general phylogenetic *trends* and not, or but rarely, of a veritable
'linear' sequence of forms belonging to an uninterrupted evolutionary
line.

LAM has repeatedly advocated a 'dynamic' approach to morphological
problems, and I wholeheartedly agree with him that the New Morphol-

ogy should, when possible, make use of data from plant anatomy, embryology, ontogeny, genetics, palynology, physiology, biochemistry, teratology and other branches of science (including elements of the classical morphology!). However, one should not be so gullible as to accept certain conclusions, reached in other disciplines, as final or unassailable. The embryological processes of double fertilisation and secondary endosperm formation have till recently generally been considered to be so unique as to render highly improbable the notion that they developed more than once during the phylogeny of the Spermatophyta. This improbability provided the principal indirect argument in favour of a monophyletic origin of the Angiosperms; but evidence from other sources (anatomy, palynology, floral biology and also taxonomy!) is indicative of a heterogeneous assemblage of not so closely related forms rather than of a homogeneous 'natural' major taxon, and indeed the idea that the Angiosperms are polyphyletic (polyrheithric) is gradually gaining ground. As I have attempted to demonstrate, the double fertilisation and subsequent development of an endosperm are not such exceptionally singular processes: they may well be the almost inevitable consequence of a general phylogenetic trend among certain advanced cycadopsid groups, so that the embryological argument has no demonstrative force and certainly does not render a monorheithric descent of the Angiosperms ineluctable.

It seems fitting to conclude this chapter with another statement of the most ardent protagonist of the modern approach to phytomorphology, H. J. LAM, viz., that the New Morphology is fundamentally a phylogenetic theory, so that any criticism should be directed in the first place at its *phylogenetic* postulates, and exclusively non-phylogenetic arguments cannot be accepted as cogent in judging its merits.

3

Definitions, Semantics and Heuristics

Morphological Definitions 'by Proxy'. Necessity of Unequivocal Semantics. Advantage of the Phylogenetic over the Typological Inquiry. Examples of Failure Through Insufficient Observance of Principles. More Fallacies.

Doch ein Begriff muss bei dem Worte sein.
J. W. Goethe, FAUST I

Clarity of definition is a major requisite in a scientific discussion of any kind, but especially in morphological disciplines, for the simple reason that morphological circumscriptions are usually comparative-descriptive rather than 'absolute' or 'quantitative'. The morphological entities to be defined are compared with other morphological structures, and morphological ('phenetic') features provide the criteria by which we relate them (*i.e.*, classify them in the same category when we assume that there is agreement in these characteristics), or distinguish them (*i.e.*, treat them as members of different categories when we think that there is some discrepancy). It cannot be helped that such judgments tend to be subjective, because the criteria are not unequivocal. Characterisation of morphological categories by means of physical or chemical standards, which might put the definition on a broader basis, has hardly been attempted. A physical or chemical approach to descriptive morphology is a hitherto neglected and promising field of inquiry. The more or less exclusive occurrence of certain chemical substances such as pigments and alkaloids in some parts of the plant body and the localisation of the sites of biosynthesis in certain organs, for instance, are strongly indicative of rather fundamental and hence useful ecological, physiological and biochemical differences between various regions of the plant. A better understanding of these and similar cases as well as of various growth phenomena, and especially of physiological and biochemical interaction between certain organ categories, may provide some useful clues to the causal forces underlying such processes as (phylogenetic) differentiation and other semophyletic (evolutionary) changes, and perhaps also throw some light on the development of parallelisms and convergencies; but the only tangible results so far achieved in this field of 'experimental morphology' are the recognition of interactions resulting in changing morphogenetic processes, notably during the induction of 'flowering' in the Angiosperms. As I shall point out in Chapter 13, the study of physiological processes and interactions enables us to make rather striking morphological deductions, but this is in its earliest stages. In any case, phytomorphological problems cannot all be solved by physiology and biochemistry alone— primarily, there must be some basic morphological framework to aid and guide the physiologist.

Being thus compelled, for lack of additional methods, to use the traditional tools of the trade, we must strive for the greatest possible efficacy to make them adequate. Frequently a rather disparaging note can be heard in statements of phytomorphologists when they have in vain attempted to find unequivocal criteria to define organ categories (such as stems, leaves, etc.), and some even feel so frustrated that they reluctantly admit that the differences ultimately 'escape' us. In his *Principia Botanica*, CROIZAT (1961) goes so far as to challenge phytomorphologists when they use such terms as 'leaves' (lateral or appendicular organs), 'stems' (axes), 'phyllomes', 'caulomes', etc., in definition, because, as he claims, they cannot be adequately circumscribed and are, therefore, of no value whatsoever in morphological argumentation. I take up the gauntlet, not only because I think things are not so bad as they are painted, but also because of the confusion by CROIZAT (and others, such as ARBER 1950) of definitions, semantics and simple heuristics. I may be an exception among phytomorphologists in that I regularly consult such unpretentious but very useful publications as excursion floras. It does not worry me in the least when in floristic literature the assimilatory organs of mosses and club mosses are referred to as 'leaves', or when fern fronds are described in terms of a dicotyledonous compound leaf as a 'leaf' with a 'petiole' and entire or dissected 'leaflets'. Every user understands what structural element of the plant is intended; and even though the *semantics* are certainly questionable from a morphological and phylogenetic point of view, the terminology is heuristically adequate and confusion is hardly possible. The trouble starts only when the reader associates with a term a mental picture different from the one the original author intended to convey.

When a morphologist mentions 'telomes', most workers understand what kind of morphological structure he has in mind, but even without a definition of a 'telome' one can discuss morphological theories in terms of telomes, one can make diagrammatic drawings or models of them, or show them 'in the flesh' in the form of fossils or parts of living plants. This is again a matter of heuristics, of transference of ideas. The most generally accepted definition of a telome is simply 'a vascularised terminal portion of a thallus', which definition one can augment by adding that a telome is usually a cylindrical organ and that the vascular tissues form a longitudinal central strand, that this strand is protostelic, that a telome has a cuticle and epidermal stomata, etc.

The reason why the term 'telome' was introduced is simply that the New Morphology requires such a name for the prototype of both axial and lateral organs, which was neither 'stem' nor 'leaf'; theoretically, one could use this phrase as the only 'definition'. The more elaborate circumscription was drawn up partly for heuristical use, because such a definition facilitates the transfer of many additional ideas incorporated in the telome theory, among them: the derivation of a telome from an unvascularised thallus after the development of conductive tissues; the occurrence in ancient rocks of telomic fossils (Psilophyta, Psilophytales) which are supposed to be the basic stock of some or of all Higher Cormophyta; and even the adaptation of early telomophytes to a terrestrial life (the stomata!). Once all this has been explained and understood, the term 'telome' conveys much more to the initiated than the simple phrase 'a vascularised terminal piece of a thallus', or 'the phylogenetic prototype of both leaves and stems', because 'there is a whole story behind it' and the semantics now include a number of concepts, postulates and deductions apart from the concise basic definition. When the term 'telome' is used in heuristics, it is not always fully realised that it is a rather complex concept; some may not 'grasp' the idea and thus may give it a somewhat different meaning. In particular, the interpretation of the morphology of the Higher Cormophyta (the Angiosperms!) in terms of telomes is in several respects an unwarranted application of the concept; for one thing, the telomes long ago lost their individuality and became incorporated in complex organs (leaves, stems, etc.) which in their turn behaved as integrated units (see also Chapter 8). The proper way to treat the new entities is to define them as 'syntelomes' and subsequently as 'leaves', 'stems', etc., each term corresponding to the evolutionary level of the organ in question. Although I should be the last to deny that a 'leaf' is the semophyletic *homologue* of a number of telomes, I do not subscribe to the glib identification of a part of a leaf as a 'telome'—it *may have been* a telome, but it is not a telome any more *as originally circumscribed*.

If after this digression we return to the subject of definition and, for the sake of simplicity, consider a short definition without frills, *e.g.*, 'A telome is a terminal vascularised piece of thallus', one might object to the original formulation because a 'thallus' and a 'vascularisation' are not defined. A trained botanist 'understands' these terms and associates a mental picture with each of them, but if necessary they can again be described, *e.g.*, a 'thallus' as a cell aggregate of a certain type and a 'vascular strand' as a tissue, *i.e.*, as a complex of cells with more or less specific characteristics (such as phloem cells, tracheids, etc.). The 'cells' (and 'tracheids', etc.) can in their turn be defined by their shape, in terms of their cell walls and their contents, or by their physical and chemical properties; but we do not do this as a rule, because such things as 'cells' are tacitly understood to represent something rather definite and are associated with a special mental picture—although everybody who has ever tried will agree that it is by no means easy and perhaps impossible to give a satisfactory definition of a 'cell'. Essentially, the *definitions* are 'translations' of the semantics into other terms, until one eventually arrives at words 'that everybody understands'. The crux of the matter is the assumption that 'everybody understands' a certain term (or definition). This also expresses the expectation that the reader (or listener) is not only sufficiently informed or 'initiated', but forms approximately the same mental picture as the writer (or speaker). Often enough this is not the case and semantics degenerate into hearsay, so that a morphological term is frequently so indiscriminately applied as to become ambiguous or even so vague that an explanation of its meaning in any particular case under discussion becomes necessary to avoid confusion. One gets the impression that phytomorphologists use various terms without themselves knowing what they are talking about, *i.e.*, without realising the implications of what they are saying or claiming, which inevitably results in circular reasoning and impermissible deductions, so that their explanations are often full of holes. This is partly caused by the typological (scholastic) approach to morphological problems, with its own semantics and heuristics, which are not necessarily the same as, nor always compatible with, those of a dynamic phytomorphology. A typologically defined 'sporophyll', for instance, differs substantially from its phylogenetic namesake.

In dynamic phytomorphology the principle of defining by reference to other terms, used in all the 'exact' sciences for the conveyance of ideas, is made possible by the phylogenetic approach. Each structure can be referred to another structure that is or was and it does not make the slightest difference if there is an element of uncertainty, of speculation, in establishing the connection; that is what a working hypothesis is for! There is no limit, because if need be we can extend our references beyond the borders of the original discipline by employing chemical and physical semantic symbols to define our terms. The typological approach, on the other hand, is scholastic because it accepts a number of entities and categories as 'given', as 'existing', and as independent or interchangeable, so that there soon comes an end to the semantics and the definitions, the ultimate reference being to the postulated categories and no further; this is precisely what CROIZAT had in mind when he declared that morphological terms (categories) could not be defined. Little wonder that the application of the Angiosperm-centred typology so often fails when it is introduced in phylogenetic botany. Its application to groups of plants other than Angiosperms (a form of applied phylogeny!) causes certain initial inconsistencies, but the difficulties and discrepancies increase enormously when typology becomes involved with fundamental phylogenetic problems. To begin with, this explains why the standard typological (phenetic) system of classification of recent taxa is essentially incompatible with any phylogenetic classification that includes both recent and extinct forms. This is a clash of fundamentals: plain taxonomy is based on typologically (phenetically) fixed entities, the taxa, but semophyletic relations require the gradual evolution of one taxon into another. Thus some fishes became amphibians, some amphibians reptiles and some reptiles mammals; similarly, Progymnosperms became Gymnosperms, some Gymnosperms became Cycadopsids and some Cycadopsids evolved into Angiosperms. Customary systematics would separate fishes from amphibians, these from reptiles, etc., and Gymnosperms from Angiosperms, but the semophyletic (historical) relation means the continuous linking of a group of mammals through its ancestral reptilian taxon to certain extinct amphibians and ultimately to an ancient group of fishes, and of some Flowering Plants, through cycadopsid, protocycadopsid and pteridospermous progenitors, to the Progymnosperms. In practice, these relations are represented in a three-dimensional diagram or 'dendrogram' showing the various putative genealogies. But any grouping along a 'horizontal' time level conflicts with any grouping along the 'vertical' time-scale and the two resulting classifications, though both meaningful, are incompatible. This is well understood by many taxonomists and they have, for greater convenience, tried to reconcile the two by some sort of compromise. Unfortunately, this is not usually so clearly understood in phytomorphological circles, although exactly the same clash of principles is found between semophyletic relationships, the 'vertical' relationships between organs and organ systems of one evolutionary lineage, and the static typological categories of organs at the present-day time level. Unlike the taxonomists, phytomorphologists cannot find a compromise without confusing other workers, or becoming entangled in the maze of their own semantics and definitions. We may merely frown upon the rather loose employment of certain morphological terms in purely descriptive floristic and taxonomic works, but in morphology and in various other aspects of phylogenetic botany one must be on guard to prevent inconsistencies and absurdities. Several examples will be found in other chapters, but a few cases will be given here to illustrate my point.

The static typological homology concept leads to an arbitrary decision as to what is homologous and what is not, whereas the dynamic homology concept postulates a common ancestry, a common archetype of homologous elements. The 'established' (!) homology of the genitalia of the Higher Cormophyta with fertile leaf homologues, or 'sporophylls', has almost consistently been accepted by the majority of the phytomorphologists (and other botanists) for at least sixty years, but there was very little phylogenetic evidence to support the wholesale transfer of the typological ideas to evolutionary botany. The sporangium is older than the leaf, and any conceivable type of fertile leaf homologue must have had a singular semophylesis, during which the sporangia and other elements (which need not yet have become leaves) came together; but this is very awkward: what would be the common prototype of the two postulated categories of lateral organs? I submit that the presence or absence of sporangia (as organs *sui generis*) would in principle deny the homology of sterile leaves with any laminose fertile organ, but one could still maintain that the two are 'appendicular' or 'lateral' (*i.e.*, non-axial) organs.

The crucial point is: When did a 'sporophyll' start its existence as a lateral organ? ZIMMERMANN believes that he has found a plausible explanation of the origin of 'sporophylls' by assuming that the sterile leaves are derived from 'sterile' telomes fused into syntelomes and subsequently differentiated into leaves, and their supposed fertile leaf homologues from 'mixed' syntelomes incorporating both 'sterile' and 'fertile' (sporangium-bearing) telomes. The homology of 'sporophylls' and 'leaves' was thus sought in a common prototype, a syntelome, which by the same evolutionary processes became a laminose organ. ZIMMERMANN's reasoning seems to be in order, both semantically and heuristically, and his suggestions can most probably be accepted to explain the semophyletic origin of the fertile fronds of some true ferns, such as those of Marattiaceae and Polypodiaceae. Taken at face value, it would seem to be equally well applicable to all those fertile organs of the Higher Cormophyta which have hitherto been considered to be 'sporophylls' on typological grounds. The fundamental mistake ZIMMERMANN made is that additional arguments in favour of the foliar nature of the genitalia of the Angiosperms were obtained from the traditional comparative morphology of recent forms, which is largely typological and usually indecisive, whereas the crucial method of inquiry is the phylogenetic one. The only point to be established is the likelihood whether indeed a semophyletic development of the reproductive organs of the Cormophyta from a mixed syntelome had occurred, for which we would in the first place require relevant palaeobotanic data. ZIMMERMANN, like so many others before and after him, refers to seed ferns with frond-born sporangia (*e.g.*, *Pecopteris*) and to *Cycas* as indications of an early development of 'sporophylls', but these isolated examples among already specialised groups are not very convincing as long as the *origin* of the fertile fronds of the seed ferns and the reproductive organs of *Cycas* remains conjectural. In some of the following chapters, the probable semophyleses of the fertile regions and the genitalia will be discussed on the assumption that the Devonian Progymnospermopsida *sensu* BECK formed the basic stock of all gymnospermous groups (and indirectly, of the Angiosperms). These ancient plants had hardly completed the change-over from the telomic stage to the more advanced cormophytic condition, but their sporangiophores were not incorporated in mixed syntelomes, and where there are no mixed syntelomes as prototypes there can be no sporophylls; in other words, the Progymnosperms were, in LAM's terminology, completely 'stachyosporous'. A careful analysis (MEEUSE 1963b) reveals that close associations between sporangia and their sporangiophores (derivatives of fertile telomes!) with foliar organs (derivatives of sterile telomes) in, *e.g.*, seed ferns are of a secondary nature and only initiated after the non-laminose fertile organs and the phyllomes, subsequent to the formation of syntelomes, had led an autonomous existence for a considerable length of time. The female reproductive organ of *Cycas* has been the subject of a lengthy discussion (MEEUSE 1963a) from which I concluded that this organ is not a 'sporophyll' by any standard or definition.

ZIMMERMANN's theory of the mixed syntelomes is simply not compatible with the available evidence and fails to convince me of the existence of 'sporophylls' among the Spermatophyta. This is an example of another pitfall: morphological working hypotheses, like those of other inductive branches of science, must hold their own when they are confronted with relevant observations.

ZIMMERMANN's hypothesis did not change the situation appreciably, and the 'sporophyll' concept kept running like a continuous thread through the 'established' morphology of the Higher Cormophyta. The vicious circle was still closed, the Angiosperms remained a group of unknown ancestry and 'dissident' suggestions such as LAM's concept of 'stachyospory' were rejected with a vengeance. In the meantime the concept of a 'sporophyll' had been corrupted to a vague notion and practically every organ bearing or associated with a sporangium or a group of sporangia has been called a 'sporophyll' at one time or other: fertile fern fronds, synangiophores of *Ginkgo*, the Coniferae, *Taxus* and the Sphenophyta, stegophylls of the Lycophyta, complex (mixed sterile-fertile) fronds of the Pteridosperms, integuments (HAGERUP 1934 *et seq.*,

CHADEFAUD 1946), the chlamys of *Gnetum*, interovular scales of *Cycade-oidea*, 'cone-scales' of the Cycadales and seed-scales of the Pinales are a fairly representative but probably not exhaustive selection. Typological considerations and definitions can be stressed very far, but they cannot be held responsible for such a chaotic assembly of heterogeneous morphological entities. The reason is simply that phytomorphologists got so completely entangled in semantics and definitions, whilst losing sight of the fundamental homologies, that only one idea stuck: there *must* always be a 'sporophyll', *i.e.*, a lateral or 'appendicular' organ (by designation, not by any other standard!). Thus, when a suitable organ could not be found, any fertile element or structure was taken to represent it even if the alleged 'sporophyll' did not even remotely fit the most elementary typological description. This circular reasoning was worsened by the retrograde interpretation of all these organs as fertile homologues of lateral organs, and some neck-breaking acrobatic stunts were performed to 'derive' the fertile organs from the postulated prototype of a 'sporophyll'.

In this connection, the erroneous application of the homology concept can be mentioned as another violation of scientific principles. The static homology concept leads to an arbitrary decision as to what is homologous and what is not, whereas the dynamic homology concept postulates a common ancestry (propinquity of origin) of homologous elements. (It is easy enough to confuse the two, as will be shown presently.) The principle of homology is also used in mathematics and this mathematical concept is the complete analogue of the neomorphological one. If the element or quantity a is homologous with the element (quantity) b $(a \sim b)$, there is a relation between a (or other elements of the category of a) and b (and all other elements of the category of b) determined by a relation of each to an element or quantity c. This I have expressed in the following symbols, which also apply to a phylogenetic homology:

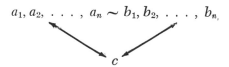

$$a_1, a_2, \ldots, a_n \sim b_1, b_2, \ldots, b_n,$$
$$c$$

There need not be an identity between any one of the elements $a_1 - a_n$ and one of the elements $b_1 - b_n$.

A typological homology is more of the nature of a mathematical near-identity: a number of elements $p_1, p_2, \ldots, p_n$ all have the same relation to one element q $(p_1 \equiv q; \; p_2 \equiv q; \; \cdots \; ; \; p_n \equiv q)$, hence $p_1 \equiv p_2 \equiv \cdots \equiv p_n$, *i.e.*, all homologues form one category of more or less identical elements.

Let us illustrate the application of the two forms of homology by a concrete case. The male reproductive organs of *Ginkgo* are sometimes developed as in Fig. 1B, Fig. 1C showing the normal condition of an axis bearing a number of synangiophores (*f.p.*). In Fig. 1B the proximal portion bears several leaf-like organs (*p.s.p.*), occasionally even a combination of *f.p.* and *s.p.* in one organ intermediate between the two types. The conventional interpretation runs as follows: the fertile organs (S) are 'replaced' by leaves (L) in exceptional cases and hence are homologous with leaves (homotopy often plays an important part in this sort of deduction), so that S must be a 'microsporophyll'. The same reasoning

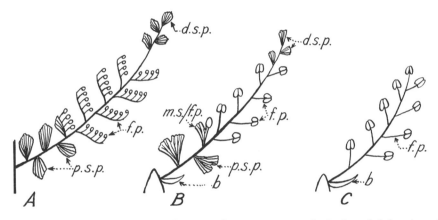

FIGURE 1. Atavism in male reproductive region of *Ginkgo biloba* (semidiagrammatic): A—Mixed 'pinna' of partly fertile 'frond' of *Archaeopteris* (*p.s.p.* = proximal sterile primules; *f.p.* = fertile pinnules; *d.s.p.* = distal sterile pinnules). B—Atavistic form of microsporangiate organ of *Ginkgo* (*m.s./f.p.* = 'mixed' sterile and fertile pinnule; *b* = bract). C—Normal male reproductive organ of *Ginkgo*.

has also been applied in a pseudo-phylogenetic sense as the result of ingrained traditionalism (*e.g.*, by NOZERAN 1955), but even if we admit that there is a fundamental phylogenetic homology between L and S it does not follow that L ≡ S at the present-day level and that S is, therefore, a *leaf*. Any phylogenetic relation between L and S must at least be reciprocal, and there is as much reason to express S in terms of L (a leaf) as L in terms of S (a synangiophore); this would make the leaf-like organ L equivalent to an axial synangiophore. I am sure that many phytomorphologists would 'feel' the last relation as an absurdity while taking the reverse ('S is the homologue of a leaf') for granted! There is good evidence that *Ginkgo* is a descendant of a progymnospermous type with a morphology resembling that of *Archaeopteris,* so that the semophylesis of the male reproductive axis of *Ginkgo* is quite clear. The condition in Fig. 1B is apparently a true atavism and represents the ancestral situation of the progymnospermous prototype (Fig. 1A), the small leaf-like organs corresponding with the proximal 'sterile pinnules' (*s.p.*) and the sporangium-bearing organs with the 'fertile pinnules' (*f.p.*) of a progymnospermous frond, so that 'transitional' forms between *L* and *S* correspond with a half-fertile–half-sterile pinnule. There is a phylogenetic relation between the organs *s.p.* and *f.p.*, but it is indirect through their common origin from telomes. The fertile pinnules *f.p.* consist of aggregates of fertile telomes (sporangiophores) which never became 'foliaceous' and S never became a leaf, let alone a 'sporophyll'; whereas each sterile pinnule *s.p.* represents a sterile syntelome and assumed the character of a leafy organ (*L*), but was never sporangium-bearing. In symbols:

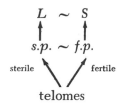

(It must be mentioned that the sporangia are organs *sui generis*, but this does not change the relations essentially, if the homology between *s.p.* and *f.p.*, and *L* and *S*, respectively, is restricted to the sterile portions of *f.p.* and S.)

The phylogenetic (dynamic) approach to the problem has several advantages over the typological method, because it gives a clear over-all picture of the relations between the sterile and the fertile organs, whereas the 'classical' interpretation can only accept or decline the homology of the fertile organ with a *leaf* (which is manifestly incorrect).

Another unwarranted deduction is the so-called *serial homology* of organs successively produced at a shoot apex. The principle of serial homology (homotypy, homodynamy) is based on the assumption that if organs develop in the same way in morphologically corresponding places (usually in a serial arrangement) they are homologous. One should bear in mind that homotypy relates organs belonging to one individual; it would seem to be of only academic interest or even slightly pedantic when, for instance, one declares that all the leaves of a single branch (which are to all intents and purposes identical, barring the normal intrinsic variation) are 'homologues'. There are, however, several examples of phenetically non-identical organs which are regarded as homodynamous and hence as homologous—the extremities of the Arthropoda and the fore- and hind-limbs of the mammals provide good examples—and the question arises as to how far one can stretch the argument. Some cases of serial homology can undoubtedly be ascribed to morphological divergence of organs derived from a common prototype, such as the fore- and hind-legs of the Tetrapoda, which can be traced to the paired fins of ancient groups of fishes, and these fins again to prototypes of a similar design (skin folds, etc.). The phylogenetic homology may be remote but is nonetheless there. However, one must be careful in applying the serial homology in cases where, in higher plants, with their 'open' type of growth, a sequence of identical organs is more or less suddenly succeeded by a number of different ones, as when a shoot apex in a 'vegetative phase' producing leaves changes into a reproductive zone, a common phenomenon during the initiation of the process generally referred to as 'flowering'. If the principle of serial homology were applicable, all organs formed by the same shoot apex—leaves, bracts, prophylls, perianth segments, genitalia—would be homologous; since this is exactly what the classical floral morphology postulates, the serial homology was accepted in this case and adduced as an additional argument in favour of the appendicular nature of the floral organs (another case of circular reasoning!), the 'transition' between petals and stamens in *Nymphaea* and other examples being used as a demonstration of the homology. My first objection is the lack of phylogenetic evidence that there is a fundamental homology between the sterile organs and the genitalia. The sporangia are organs *sui generis,* and organs bearing sporangia must therefore always be of independent origin in respect of sterile organs; in fact, the simple presence of sporangia indicates that the genitalia cannot be complete homologues of vegetative leaves. An important difference between the vegetative leaves and the genitalia of many Dicotyledons is the presence of an axillary bud in the vegetative region and the absence of a bud in the axils of the genitalia, which either points to a fundamental difference between leaf and reproductive organ or to the equivalence of the genitalia with a leaf *plus its axillary bud* (meaning that the genitalia contain an axial element); I do not accept the traditional explanation that the genitalia have simply 'lost' their axillary buds (which is again circular reasoning!). The histogenesis of the vegetative organs often differs from the early development of fertile organs derived from 'the same' shoot apex in its reproductive phase (see, *e.g.,* PANKOW 1962), and this again is indicative rather of fundamental differences than of homology; even the morphology of the shoot apex itself often changes appreciably when it enters its reproductive phase. The physiology of 'flowering' strongly suggests that certain physiological and biochemical processes result in a different morphogenesis at a shoot apex when 'flowering' is induced, which again pleads against the homology of fertile organs and vegetative organs. The sequence of vegetative phyllomes and floral appendages produced by a single shoot apex is apparently not a case of homodynamy and does not warrant the conclusion that all the organs of the sequence are (serial) homologues.

As a final example, the almost notorious subject of teratology. There has been a tendency to explain all sorts of abnormal developments as 'throw-backs' (atavisms), *i.e.*, as indicators of ancestral conditions. In a very lucid paper, HESLOP-HARRISON (1952) has pointed out that there are in fact two forms of abnormalities, *viz.*, (1) all deformations and malformations caused by irregularities in growth processes and by other unusual conditions which throw the regulating mechanism of morphogenetic development out of balance, abnormalities which are of course of little direct consequence from the phylogenetic point of view, and (2) the genetically controlled true atavisms representing ancestral stages (which are usually not deformed but may be more or less rudimentary). Antholysis (virescence, phyllody, etc.) clearly belongs in the first category, so that morphological deductions based on abnormally developed flowers of this type cannot be relied upon to establish (phylogenetic) homologies, in other words, phyllody of 'carpels' and stamens does not 'prove' that the genitalia are leaf homologues. The examples of the second category given by HESLOP-HARRISON include such cases as the development of a fifth stamen or a supernumerary staminode in normally tetrastaminate zygomorphic flowers of Scrophulariaceae (reflecting the probable origin of such flowers from a pentamerous actinomorphic prototype resembling, *e.g.*, *Verbascum*), but as we have seen (see p. 24) the abnormal development of the reproductive organs of *Ginkgo* provides an even more instructive illustration of truly atavistic teratologies and of their value as indicators of ancestral conditions.

When, to sum up, we strike a balance, the picture is not too bright. Morphological concepts and definitions got watered down and corrupted, so that much of the vocabulary became more a matter of hearsay than of a clear understanding of principles. Morphological deductions are often based on false assumptions or, what is worse, the reasoning often seems to be aimed at saving an 'established' postulate or working hypothesis at all cost. Antiquated theories came to be regarded as 'proved' and proclaimed in textbooks and manuals as factual knowledge, as gospel truth, but the traditional main-stays of the Old Morphology do not appear to be built on sound foundations. The clashing of the different fundamentals of the static and the dynamic approach of morphological problems was not even suspected, so that false (non-phylogenetic) homologies and typological semantics were indiscriminately applied in so-called phylogenetic inquiries. Apart from attempting to lay bare the inconsistencies and absurdities, I can do no better but quote an old saying: *Errare humanum est, perseverare diabolicum.*

4

Phytomorphological Schools and Morphological Traditions

Morphological Schools. Morphological Theories. The 'Type' Concept. Homologies and Pseudo-Homologies. The Need for a Synthesis.

'Authorities', 'disciples', and 'schools' are the curse of science; and do more to interfere with the work of the scientific spirit than all its enemies.

The quotation above, attributed to T. H. Huxley, is probably more applicable to phytomorphology than to any other biological discipline. It is only fair to acknowledge that the establishment of a 'school' has often been beneficial, in that many scientists have received at least some of their training in a renowned centre of learning and subsequently disseminated the acquired knowledge. For various reasons the influence of such centres upon the development of a branch of science has been greater in the past; but, despite the specialization which is today the usual source of reputation for centres of learning, this process of clustering at such places, followed by dissemination of the acquired scientific experience, continues on a smaller scale to this day. Workers wholly or partly educated in the same centre tend to have a common field of interest, to employ the same methods of research and to adhere to similar if not identical theories—they form a 'school'.

Detrimental consequences of such centres and 'schools' can be a dearth of fresh thought resulting from adherence to ideas established so long ago that they are more than obsolete, the ignoring of the findings of competitive schools and the negation of new and relevant scientific information, especially if the static and conservative attitude is coupled with an exaggerated admiration or even veneration of the personality and the teachings of an authority, of the 'master', so that even a suggestion of criticism is anathema. Even if a 'school' is not adverse to renovation,

its scope may remain too narrow if scientific discoveries in related disciplines are not integrated in its own working hypotheses. Either case
leads to an isolation of the phytomorphological school and, inevitably,
to its progress' lagging behind that of other branches of botany. How
bad things may become is probably best demonstrable by TROLL's
deliberate restriction of his field of study to 'Gestalt' morphology, which,
as we shall see, is essentially still the same approach as that of the old
idealistic morphology. The Trollian morphology is scholastic in its
approach and represents in modern science an anachronism. This is
bad enough, but its narrow basis, the wilful limitation to typology
('Gestalt') leads to a rigidly static and one-sided method of inquiry. A
confrontation with other methods of research is thus precluded; and although the conclusions of such 'classical' typological studies may impress
other typologists, they are not always acceptable to the majority of the
neomorphologists, simply because any scientific conclusion, every working hypothesis and theory, must hold its own when confronted with evidence from whatever source. Morphology is the scientific study of the
kinds and diversity of constituting elements of organisms and of any
and all relationships among them. Typology, the comparison of 'Gestalt'
features, is only one of its aspects and a very poor indication of relationship at that. When adherents of such self-contained morphological
schools explain that 'one has to acquire that special morphological feeling' one wonders what requirements the uninitiated must fulfil to become
eligible for membership in such a secluded and mysterious society.

Other phytomorphologists and phylogenetic botanists also tend to be
conservative and one-track minded, though not so ostentatiously. Both
TAKHTAJAN in his *Evolution der Angiospermen* (German edition, 1959b)
and EAMES in his *Morphology of the Angiosperms* (1961), whilst adduc-
ing corroborative evidence in their argumentations, minimize or mis-
represent contradictory findings and often dismiss dissident opinions and
alternative theories with a summary self-assurance that verges on contempt. They write with so much self-assumed authority that the inex-
perienced student may be completely overawed and is likely to accept
their apodictic statements as 'the last word' in a particular branch of
botany. Adoption of scientific principles by accepting the teachings of
an 'authority' was a medieval ('scholastic') tradition—'*Galenus dixit*'—it
is certainly not promotive of educating a budding scientific mind towards
independent thinking. More freedom of thought is undoubtedly much
needed in phytomorphology and this is, among other things, my apology
in anticipation of any criticism that the number of preliminary and general chapters appears unduly high.

Another source of a babel of tongues and of segregational tendencies is methodology. There are 'classical' (typological), 'anatomical',

'ontogenetic' (morphogenetic) and 'phylogenetic' schools, each with its own postulates, its own terminology and its 'own' literature. These 'schools' tend to widen the gaps between themselves and the other ones by a kind of self-imposed (and, in their own eyes, splendid?) isolationism. They consistently use the same train of thought, the same arguments pro and con, and they often cite the same (mostly their 'own') group of authors over and over again, dismissing others as irrelevant. Each school runs the risk of becoming a select and secluded community, fully comparable with a religious sect that can only find a responsible audience among its own adepts but cannot convince non-believers. I cannot follow PURI (1960, p. 107) when he states that alternative morphological concepts and interpretations can co-exist.* I have pointed out already (see Chapter 1) that one can build up a working hypothesis on a variety of postulates and that this is not unscientific; but when a number of alternative theories exist, one should not accept this with resignation but rather strive for a synthesis. This is the only way to make significant advances. One must never forget that non-morphologists are inclined to look down upon morphology as an old-fashioned, static and rather incomprehensible (read: *muddled*) branch of biology. When it comes to the application of morphology in, *e.g.*, phylogenetic considerations, which concepts and which working hypotheses must the non-morphologist adopt? I have repeatedly demonstrated that certain conventional ideas have stood in the way of scientific progress and have most likely prevented an earlier solution of such problems as the origin of the Angiosperms. OZENDA (1946), LAM (1948 *et seq.*), NOZERAN (1955), PANKOW (1962) and others have pointed out that the time is ripe for a synthesis and that serious attempts must be made to reconcile the various viewpoints. I would go one step further and say that a synthesis is long overdue. The difficulty is that though there are several morphological traditions, a number of notions and hypotheses are rather generally accepted as factual by a large majority, or at least by the protagonists of one school, and that such 'established' (traditional) concepts die hard. The 'sporophyll' concept is one of them. As I have pointed out before (MEEUSE 1962), such conventional points of view are indiscriminately applied to theories based on different fundamentals, which is scientifically illicit. An additional complication of such a transplantation of incongruous phytomorphological concepts is that this is sometimes done with and sometimes without restriction, redefinition or other modification of their original circumscription. An example of the

---

* This is not at variance with the statement on p. 5 concerning different approaches to a problem, which may result in alternative working hypotheses. Progress in science means that only one theory is retained and other ones rejected after a thorough confrontation of all theories with the accumulated evidence.

introduction of borrowed ideas is the application of static typological concepts to phylogenetic botany. All this is very confusing; frequently one can read and understand a phytomorphological publication only if one knows, or if one can sense, to what school its author belongs. The best, but a laborious way to sort out the muddle is to trace every notion, every principle, every term and every hypothesis to its origin and to follow its history in the various schools. I gratefully acknowledge the painstaking spade-work done by my collaborator B. M. MOELIONO in unearthing various sources and in pin-pointing the evolution and mutation of various concepts. MOELIONO's conclusion, that the writings of some frequently cited morphologists are often misquoted or misrepresented, also sets one thinking. One should certainly strive for clarity of definition; SIMPSON's recently published *Principles of Animal Taxonomy* (1961) is exemplary in this respect. In the following survey, I shall attempt to give a brief historical sketch of the development of several schools and their fundamentals.

Morphology is essentially the study of 'form' in all its aspects, and an analysis of shapes and structures can be done only by comparing the forms with one another or with a common standard, or both. The guiding principle is of course that the more alike the forms, the more closely are they 'related' to each other—related in the sense of *any* common binding principle. The relationship may also be a correspondence in the *origin* of the form either through a similar morphogenetic (onto-genetic) development or, phylogenetically, through propinquity of ancestry. The recognition of 'degrees' of correspondence leads, as it does in taxonomy, to a classification of forms into categories, the categories being distinguished by smaller or larger discontinuities in visible and measurable characteristics or in origin. In the early days of scientific botany, a 'plant' was practically synonymous with what we call a 'higher' plant nowadays, in fact with an Angiosperm, so that phyto-morphology began as the morphology of the Flowering Plants and was thus originally 'Angiosperm-centered'. Since time immemorial, various categories of essential parts of these plants had been intuitively recognised, and had been expressed in the vernacular by such terms as stem, leaf, root, flower, fruit and seed. The first morphologists had simply grown up with these notions and accepted these categories as discrete (*i.e.*, as definite or invariable) entities and as alternatives (*i.e.*, a given part of a plant could belong only to one group of organs and not to another). The first problem they were confronted with was the question of what binding principle holds all parts (organs) of one category together. In pre-Darwinian times this binding principle, *homology*, was accepted as a primary and inherent characteristic of a morphological category in the sense that homologous entities were primarily defined as elements *belonging* to an existing category, rather than as elements having something in common so that they can be grouped together and *form* a category. This is in fact the basic difference between the aprioristic *typological* approach (elements belonging to one category are comparable) and other morphological methods of inquiry (elements having something in common are connected by some collective binding principle and consequently can be united into one group which is different from any other group of elements lacking that binding principle).

The first theoretical consideration regarding equality of organs by C. F. von Wolff (1759) was based on morphogenesis ('ontogenesis'), *i.e.*, on what was later termed *serial homology*. From the repetition of organ formation at a shoot apex, Wolff concluded that all organs formed in this identical way are 'members of one family'. It is of course true that identity of origin is a binding principle and a criterion of homology, so that serial homology still forms an important argument in morphological theories today, but, as I have pointed out elsewhere (Meeuse 1963c), the deduction of homology from an identical ontogeny rests solely on identity of development, so that organs formed successively in a certain area are homologous only if the processes of morphogenesis remain constant. When the morphogenetic processes change (as is, for instance, the case when a vegetative shoot apex becomes transformed into a pre-floral apex or transition apex at the onset of flowering in many Angiosperms), the organs formed before the change and those formed after the change are not necessarily homologous (see also Chapter 13).

In the Goethean morphology, the homology concept and the occurrence of certain categories of organs were expressed in the postulate of the 'idea', essentially a Platonic philosophical doctrine. The binding principle of all homologous organs was the common 'idea', later replaced by other terms with almost the same meaning, such as 'basic pattern', the (morphological) 'type' and the (abstract) common 'Gestalt'. All organs that are homologous were supposed to be the expressions of a single 'idea', somewhat of the nature of a mental blueprint, pattern or design; in pre-evolutionary times, in fact, they were often thought of as the 'idea' a Higher Being had in mind when he created numerous different forms. Each actual (living) organism and every organ was the materialisation of an 'idea'; homologous organs (and 'related' organisms) were comparable to many variations of the same musical theme and each individual expression of the 'idea' was more or less different from the other ones; but all homologous organs bore a certain resemblance to the basic pattern. The homology concept was, therefore, preconceived (postulated); nor are matters altered in the slightest if the term 'idea' is replaced by the (morphological) term 'type' or by 'Gestalt', even if the common basis of comparison (the 'pattern') is not an abstraction but an existing form which is taken as the fundamental example to which all its homologues can be related. The contemporary version, TROLL's 'Gestalt' morphology, is still based on the preconceived homology concept, but many other workers have applied the phenetic homology concept without sufficiently realising its dogmatic character.

It cannot be denied that this typological approach has produced important results. Typology has enabled the nineteenth century botanists to lay the foundations of taxonomic classifications of considerable practical value. It also made it possible to establish some sound homologies, such as the interpretation of a spine or a tendril as the morphological equivalent of a leaf, a stipule, a petiole, a branch (stem), or a peduncle of an inflorescence. In this case it is understandable that the spine (or the tendril) is regarded as the 'derived' or 'advanced' stage and the leaf (or the other organ) as the primitive (original) condition, because spines or tendrils do not resemble the average type of leaf (or stem, etc.) and in terms of the idealistic morphology this means that they show the least resemblance to the 'idea' (the 'type') of a leaf. Later, the typological appreciation of the degree of resemblance (or of dissimilarity) was rather uncritically translated into the phylogenetic meaning of 'primitive' and 'advanced', *i.e.*, as earlier, more ancestral forms or characters against the more recent ones descended or secondarily developed from them. The most unfortunate mistake made was to accept *typological* conclusions ('derivations') and classifications *as if* they had been the result of evolutionary sequences. Series of typologically homologous organs can sometimes be read in one direction only, as in the sequence of leaf to spine (or tendril), because nobody would query the primitive condition of the leaf and the derived status of the spine or tendril. However, the direction of a homologous series is not always so clear. The morphology of the Angiosperm fruit offers an illustrative example. Must the series from primitive to advanced be read in the direction fleshy to dry, dehiscent to indehiscent, loculicidal to septicidal, or *vice versa?* The typological solution is simply to recognise several alternative 'types' of fruits (*e.g.*, a 'fleshy' series and a 'dry' one, a dehiscent series and an indehiscent one, etc.), but this is evading the issue. One can also, as TROLL has done, postulate a 'composite' basic 'type' of ovary that combines apo-, para- and syncarpous portions with marginal, 'axile' and laminal placentation, but this is idealistic morphology at its worst.

It is clear that the typological appreciation of 'primitive' or 'advanced' conditions is purely subjective, the degree of phenetic resemblance to the abstract or postulated 'type' being the yardstick. The wholesale assimilation of established typological points of view in phylogenetic working hypotheses has been an understandable but most deplorable scientific blunder, from the aftermath of which we are still suffering to-day. As can be expected, the pre-evolutionistic homology concept often cuts through our present-day notions of phylogenetic relationships. If the idea 'leaf' is expressed in the form of a Lyco-leaf, a fern frond, a 'needle' of a Conifer and an assimilatory organ of a Cycad or an Angiosperm, all these organs are homologous in the scholastic sense, but most contemporary phytomorphologists do not accept a true homology of all these functional 'leaves'. Another example is found in the old-fashioned question whether a coniferous 'cone' is a 'flower' or an 'inflorescence' (which is, incidentally, also an illustrative example of the traditional, Angiosperm-centred terminology). Up to this day such associations are, perhaps subconsciously, still part of phytomorphological thinking because we so often apply typological inquiries to taxonomic and phylogenetic problems. The term 'sporophyll', for instance, has been applied to various associations of sporangia with a supporting or otherwise associated organ: to a Lycopsid 'bract' (stegophyll) with its adnate or axillary sporangium; to the peltate synangial complexes in *Equisetum* and the male *Taxus;* to a fertile fern frond; to an integument (HAGERUP); to the female reproductive organs of the Caytoniales and of *Cycas;* to the ovuliferous scale in the Coniferales (also by FLORIN, who clearly demonstrated its origin from a brachyblast, an *axial* organ!); and to stamens, to 'carpels' and to several other fertile organ or organ complexes. If some or all of these kinds of 'sporophylls' are considered to be homologous (as some phytomorphologists seem to take for granted), this is pure typology, the 'sporophyll' concept being the primary supposition which is subsequently applied to various categories of phylogenetically non-homologous organs.

Similarly, typological classifications of the Angiosperms were taken for granted by botanists and accepted as phylogenetic 'systems', but they completely overlooked the fact that the typological method started from the fundamental postulates that (1) the Angiosperms belong to one category, homologues of the 'type' Angiosperm, (2) all organs of the Angiosperms are homologous, so that (3) all Angiosperms and all their organs form homologous series and can, accordingly, be compared. The more or less 'successful' squeezing of the organs of the Angiosperms into a number of categories, into flower diagrams and into diagnostic descriptions of taxonomic units, is not an ultimate and final achievement, but only the result of the consequent applications of the basic postulates. If, like the present author, one does not subscribe to a monophyletic development of the Angiosperms, the fundamental homology of all stamens, of all ovaries and of all 'flowers' is not a foregone conclusion and, in fact, improbable. Traditional homologies become pseudo-homologies, typological similarities often become convergencies or parallelisms, supposedly phylogenetic relationships become again what they had always been—pseudo-phylogenies.

In the past, much importance was attributed to the 'comparative' typological method, and its application is still advocated by contemporary workers. EAMES quotes in the preface of his own work, *Morphology of the Angiosperms* (1961), the words appearing as a maxim at the beginning of VELENOVSKY's textbook (*Vergleichende Morphologie*) of 1905: '*Zur morphologischen* Lösung *werden wir—wie immer—die vergleichende Methode in Anwendung bringen. Auf diesem Wege werden wir zu der* richtigen *Ausschauung gelangen*' (my emphasis), and adds that he 'agrees heartily with this point of view'. I am not the only one who does not accept an unequivocal '*Lösung*' of the problems nor the '*Richtigkeit*' of these views. The modern comparative morphology is of course—at least theoretically—based on the sound principle of the homologous series of organs derived from a common phylogenetic prototype, but it must accordingly have phylogenetic (palaeobotanic) support A comparative morphology based on the static morphology of recent forms is nothing but old-fashioned typology in disguise, so that it tacitly postulates the categories, *i.e.*, the homologous series, first and leads to circular reasoning just as much as the typo-morphological systems of classification. The static comparative analysis also fails in that the recognition of 'primitive' and 'derived' conditions is entirely preconceived. The conventional comparative method treats all Angiosperm ovaries, for example, as derivatives of 'carpels' or assemblages of such 'carpels', the ovaries of the Magnoliales (or 'Ranalians', or 'Polycarpicae') being assumed to represent a very primitive condition; from this (or from a very similar) type, the assumption continues, all other forms of ovaries are derived, *e.g.*, the chlamydote Piperaceous pistils. To my mind the gynoecia of the Piperaceae are of a more ancient type, but the more recent ovary type of the Magnoliales is *not* the derivative of an ovary of the Piperaceae. A carpel of the Polycarpicae is a modification of a *truss* of essentially arillate ovules, *each* homologous with an ovary of the Piperaceous type which, to my mind, is nothing but a *single* chlamydote ovule (in fact, a one-ovuled homologue of a cupule). The *ovules* of the Polycarpicae can be compared with the homologous pistils of the Piperaceae, but not the Ranalian carpels. The traditional homology of all ovaries (or carpels) is false in that *several* non-homologous categories of pistils occur among the Angiosperms, so that a comparative analysis of the morphologically heterogeneous gynoecia of *all* Angiosperms is utterly inane and the derivation of a Piperaceous type of 'ovary' from that of a *Magnolia* equally absurd.

One can understand that, when the theory of evolution was gaining ground, morphologists were inclined to retain the established typological notions and achievements, the more so because the fossil evidence was hopelessly inadequate and phylogenetic relationships had to be deduced by inference. However, it is high time that we admit that a serious mistake was made when typo-morphological deductions were identified with phylogenetic happenings. Botany was endowed with a typological inheritance that works quite satisfactorily in such branches as applied taxonomy (herbarium systematics), but stood in the way of the progress of morphology and phylogenetic botany. It would be sanctimonious to deny that, even if one aims at a phylogenetic interpretation of homologous series, at the recognition of evolutionary sequences during organ phylogenies or *semophyleses*, the basis is still a typological approach. There is a fundamental difference, however. The typological prototypes, the primitive conditions and the series of derivatives are entirely preconceived, whereas the palaeobotanic data provide at least some tangible evidence of certain sequences in the evolution of taxa and of their organs. In fact, the traditional, *i.e.*, essentially typological, ideas have *hampered* the recognition of fossil records and of their proper place in evolutionary sequences. This was especially the case in the much debated and seemingly mysterious ancestry of the Angiosperms (see MEEUSE 1961a, 1962; also Chapter 20). VAN TIEGHEM and other phytomorphologists (*e.g.*, PURI 1951, EAMES 1961) have repeatedly claimed that anatomical and ontogenetic evidence bears out morphological (typological!) conclusions, but this is wishful thinking. They have only shown, with a greater or smaller amount of success, that certain anatomical and ontogenetic features can be *interpreted* in the light of the Old Morphology, but theirs is not the only possible and unequivocal interpretation. Only if anatomical characteristics and trends are supported by phylogenetic evidence can they be accepted as relevant. When EAMES (among others) uses the double trace leading to the placental region of a carpel as evidence of the appendicular (foliar) origin of the Angiosperm gynoecia, the whole argument rests on the occurrence of a double or treble leaf-trace in some vegetative phyllomes in some Angiosperms which is accepted (postulated!) as the primitive condition. I have evidence that in some Polycarpicae (OZENDA 1948) and Monocotyledons numerous vascular strands enter the leaf, and there is also evidence that some 'ovaries' are derivatives of pteridospermous cupules, *i.e.*, of chlamydote ovules (see Chapter 15) and hence cannot be foliar organs; thus the double placental trace can at any rate not be regarded as a cogent argument in favour of the appendicular origin of *all* angiospermous gynoecia (see also MEEUSE 1964b). The occurrence of 'inverted' vascular strands in Angiosperm gynoecia, another argument often brought forward in support of their appendicular (foliar) nature, can be explained by alternative and equally plausible assumptions (MAJUMDAR 1956). The anatomical data, therefore, do not have much demonstrative force and the whole argument based on anatomical evidence is, properly speaking, another case of circular reasoning—the homology of 'carpels' and 'leaves' is preconceived and a particular (forced!) interpretation of anatomical structures as specific features of appendicular organs is meaningless.

Another important morphological school of thought is formed by those workers who deny the homology of certain organ categories with other categories which in the classical morphology are considered to be their equivalents. The point at issue is the interpretation of the genitalia of the Angiosperms. In the majority of the phytomorphological theories, the genitalia are supposed to be the homologues of 'foliar' (lateral, appendicular) or of 'axial' organs, or of combinations of elements belonging to both categories of sterile organs, but some morphologists do not accept this homology and regard the genitalia as organs *sui generis, i.e.,* as organs of completely independent origin. Their arguments are mostly deduced from developmental studies; and, indeed, a different mode of morphogenetic ('ontogenetic') development of the genitalia as compared to the vegetative organs is frequently observed. This is particularly striking in those cases in which a vegetative shoot apex producing functional (vegetative) leaves changes into a generative apex during the initiation of the processes commonly referred to as 'flowering'. There is a difference in organogenetic development and in morphology between the leaves and the 'floral appendages' (especially the genitalia), which is undoubtedly indicative of a discontinuity in the serial formation of the organs at the shoot apex; the plausible inference is that there is no continuous serial homology between the vegetative organs and the fertile ones. Up to this point I can follow the train of thought of the protagonists of the *sui generis* theories and I also accept the *sui generis* origin of the sporangia (and their homologues such as the nucelli of the ovules), but I do not agree with the conclusion that the genitalia as a whole are *sui generis.* The observations and argumentations of the *sui generis* schools of thought are very important, because they have made us aware of certain differences in organogenesis between vegetative and generative zones which cannot be explained in simple terms of 'stems' and 'leaves' (of lateral and axial organs) and can certainly not be summarily dismissed as unimportant and irrelevant. However, protagonists of the *sui generis* origin of the genitalia have neglected the phylogenetic aspect—they base their conclusions almost exclusively on developmental studies—and also disregard the possible interaction of organs. If the *interaction* (which can be accomplished by phytohormonal substances produced in one organ and exerting their action in another one after having been transported to the reactive region) more or less suddenly changes, for instance as the result of a difference in the nature and the quantity of hormonal principles received by the reactive organ, differences in morphogenesis are likely to result. This possibility throws a different light on the *sui generis* interpretation (as will be more amply shown in Chapter 14), but is by no means a vindication of the traditional static morphology.

The ideas of protagonists of a New Morphology have been mentioned in a previous chapter and need not be repeated here. The differences of opinion and the various interpretations require an entirely new approach based on a synthetic treatment.

Morphology being the study of 'form' in all its aspects, one cannot evade the necessity of comparing forms and relating them to one another or to a common basic form. Even if one accepts certain organs as *sui generis* as opposed to others, and recognises fundamentally different categories, the constituting elements of one category must be studied in relation to each other and to elements belonging to other groups of organs. We have thus returned to the starting point and must discuss the binding principles uniting the members of one category whilst distinguishing them from the other categories. This homology concept will be discussed in the following chapter.

5

Intricacies of the
Homology Concept

Homology Concepts. Homology Versus Analogy. Convergencies
and Parallelisms. Degrees or Forms of Homology. Various Practical Difficulties.

As we have seen, the homology concept is much older than the
theory of evolution. The categories of organs were defined on intuition
(postulated) and so was the communal basic form of each distinctive
group of homologous organs, the 'idea', 'pattern', morphological 'type' or
common 'Gestalt'. The 'type' could also be a more concrete object such
as a drawing, a diagram, a hypothetical reconstruction, or even an existing form. Typology in this form is still customary; in many textbooks
'theoretical' diagrams are given from which extant forms are 'derived' (a
good example is the theoretical diagram of the flower of the Orchidales,
in which two trimerous whorls of stamens are indicated, the four or five
stamens that have 'disappeared' being indicated by crosses) and, frequently, supposedly homologous organs of a number of recent taxa are
aligned in a series, the sequence being assumed to represent the successive stages from primitive to advanced (*e.g.*, the well-known series of
so-called 'megasporophylls' of the Cycadales, leading from *Cycas* via
Dioon, Ceratozamia, etc., to *Encephalartos* or *Zamia* and suggesting the
derivation of all forms from a prototype corresponding with the condition in *Cycas;* this example is particularly suited to demonstrate the
weaknesses of typological methods, as I shall presently point out).
After the introduction of evolutional ideas in biology, the postulated
typological prototypes were replaced by the phylogenetic prototypes or
the ancestral forms or conditions, and thus the homologous series from
primitive to advanced became the complete parallel of the evolutionary
history of the taxa from more ancient to more recent forms. The change-
over to the evolutionary homology concept was deceptively easy; even the
terminology ('primitive' and 'advanced' conditions) could be adopted.

The danger lay, and still lies, in uncritical acceptance of typological findings, as if the latter indeed always reflect the phylogenetic processes and sequences.

We dare not hope ever to be able to 'prove' the existence in the past of a continuous phylogenetic lineage, except in a few exceptional cases in which fossiliferous strata of considerable thickness, representing a more or less continuous deposit and hence the evolutionary history of flora and fauna over a long period, supply continuous sequences of gradually changing forms. All we can do is to postulate a phylogenetic genealogy, using all available evidence, and build up the evolutionary sequences in the phylogeny of the organs, the *semophyleses*, along our framework.

The modern homology concept can, accordingly, quite easily and unequivocally be circumscribed, *viz.*, as the phylogenetic relation of a certain element belonging to a morphological category of organs or organ complexes, or of a particular feature of such an element (a special kind of cell or tissue, such as a microspore, an epidermal cell, a stoma, secondary xylem, etc., or even the characteristic structure or sculpture of a cell wall and any other feature of diagnostic value), to corresponding elements belonging to the same lineage. A more concise definition of homologous features is 'all elements forming a single uninterrupted semophylesis', provided one extends the meaning of the term 'semophylesis' to include the phylogenetic history of all characteristics of diagnostic value, such as the pitting of a tracheidal cell wall or the sculpture of a pollen grain. Provided one can adequately describe and delimit an organ or any other diagnostic feature, the phylogenetic history of this element can be followed in its course, but in actual practice one often works 'backwards' and retraces the changes in the element in order to establish the most probable starting point of a semophylesis. Semophyleses providing the building stones of the phylogenetic taxon genealogies and *vice versa*, one falls back on typological methods and to circular reasoning, so that we should not beat about the bush but frankly admit that we postulate ('accept') a certain evolutionary line. There is, however, a sound source of guidance for our intuition, *viz.*, the palaeobotanic evidence. There are many examples of organs which, on the ground of logical deductions combined with information provided by the fossil records, simply *must* be derived from certain prototypes and not *vice versa*. An ovule is a complex structure which contains the homologue of a megasporangium and the megasporangium is undoubtedly the more ancient condition; siphonogamy is a more derived (advanced) phase than zoidiogamy; wood vessels represent a later phase of the evolution of the elements of the secondary xylem than tracheids; Gymnosperms are 'older' than the Angiosperms; etc. Consequently, certain typological homologous series can be read in one direction only; this is a step in the right direction.

The phylogenetic development of a group of organisms can be diagrammatically represented by a number of minor phylogenies (lineages) that can be retraced to a smaller number of such phylogenetic lines and these again to a few only or, if the group is monophyletic (monorheithric), even to a single ancient phylogeny. We thus arrive at the 'genealogical tree', or *dendrogram,* as TAKHTAJAN (1959a, 1959b) prefers to call it. The above-mentioned phylogenetic homology concept can be applied to such a branched system of evolutionary lines and, starting from each of the ultimate ramifications at the present-day level, one can retrace the semophyleses to a more ancient common line of descent and, in monophyletic groups, to an ultimate base on the common 'trunk', representing the common ancestral (or 'basic') taxon of the whole group of recent forms represented by the dendrogram. Barring complete 'loss' of an organ by complete reduction, there is a semophylesis retraceable from every ultimate ramification to the corresponding organ of the basic taxon. Just as the phylogenetic lineages sprout from the same 'trunk', so these semophyleses have a common origin, and they have a common binding principle which unites them and distinguishes them from any other group of semophyleses based not on the same basic taxon but on a different 'trunk'. This is a second form of homology, the 'indirect' homology of organs belonging to different (parallel) lines of descent *through common origin* (see Figure 2). Most workers would consider such organs belonging to parallel evolutionary lines as homologues, at least if the common origin is clear and the 'divergence', the splitting up of the basic ancestral taxon, does not lie so far back in the geological past that the common origin becomes obscure. The possible homology of organs belonging to semophyleses which have had such a long independent existence that morphologists are inclined to regard similarities between organs as analogies or convergencies tempted LAM to make the somewhat paradoxical statement that 'every analogy is fundamentally a homology'. Indeed, one could retrace the phylogenetic development *ad infinitum et ad absurdum:* for instance, all Cormophyta to a telomic common ancestral group, subsequently to a thallophytic one, to a unicellular organism and, ultimately, to the pre-cellular stage of organic evolution. I doubt that all phytomorphologists have understood that LAM intended only to emphasise that homology can be relative, that there are degrees of homology. There are several cases in which the same functional adaptation of an organ occurred in independent phylogenetic lines, as in the development of an 'aril' in *Taxus* and in the Angiosperms, the cladodic development of the stems in the coniferous *Phyllocladus* and in several orders of the Flowering Plants, the repetition of the ecologically adapted juncoid, spartioid and ericoid habits, the heterospory in Lycophyta, Sphenophyta and Spermatophyta, winged diaspores in various groups, etc. In other respects as well the homologies cannot be indefinitely extrapolated into the past. I have explained in a previous chapter that, although it is feasible to explain the morphology of the Higher Cormophyta

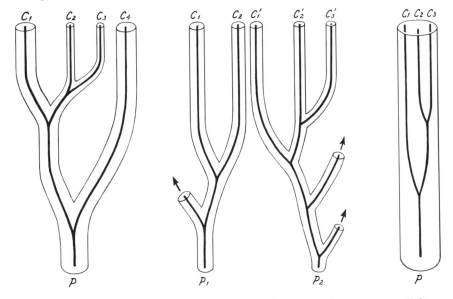

FIGURE 2. Different forms of (phylogenetic) relationships and parallelism: *Left*—A phylogenetic lineage (P) splitting into derivative lines, the semophyleses of the homologous expressions of a character $(C_1, C_2, C_3,$ and $C_4)$ based on a common archetype. *Centre*—Two phylogenetic lineages $(P_1$ and $P_2)$ as in the previous figure, the phenetic characters C_2 and C_1 'converging'; this may be an analogy or a parallelism, depending on the relationships between P_1 and P_2. *Right*—A phylogenetic lineage (P) in which a character shows a divergent semophyletic development into three different phenetic features, C_1, C_2, and C_3, which are homologous.

in terms of the telome theory, this interpretation does not take into account the additional 'acquired' characters which make, for instance, a leaf more than just a fused (webbed) flat ramified bundle of telomes. There are changes in symmetry, in polarity, in function, in its physiological and biochemical nature and in its histogenetic organogeny, whilst mutual relationships and correlations with other organs develop, *e.g.*, between lateral and axial organs, or between assimilatory organs and the shoot apex during flowering. In this way, gradually, new categories of organs (*e.g.*, stems and leaves) increasingly divergent from the organs found in the common ancestral stock (*e.g.*, telomes), develop as two or more parallel semophyleses in the same phylogenetic line or geno-rheithrum. An illustrative example is provided by the evolution of the sporogenic organs in the Spermatophyta. The isosporous sporangia became heterosporous and, from that time forward, the semophylesis of the isosporous sporangium became divided into two independent and parallel semophyleses in every individual lineage, the one leading via microsporangia to the loculi of the thecae of the stamens, the other via the megasporangium ultimately to the nucellus of the angiospermous ovule. There are, accordingly, two ways of divergence of an organ (or a character), a divergence into two (or more) semophyleses each belonging to different phylogenetic lines, and a divergence into two (or more) semophyleses belonging to a single (the same) lineage. This last phylogenetic relation represents another (a second) form of indirect homology through common origin. It can be illustrated by the development of different categories of phyllomes, such as cotyledons, functional leaves (assimilatory organs), bud and rhizome scales (cataphylls, etc.), various kinds of bracts, prophylls and other fundamentally sterile lateral organs of the fertile region. They are probably all derivatives of a uniform type of phyllome, *viz.*, of the functional leaf of the sporophyte of an early Cormophyte that developed 'pteridophytically', *i.e.*, from a free-living gametophyte. When the first developmental stage of the sporophyte was retained within the megaspore on the mother plant in the form of an embryo, the first leaves became the cotyledons, which acquired certain characteristics that distinguished them from the remainder of the leaves (and became storage organs). Similarly, bud-scales, bracts, etc., became segregated as special categories, so that, finally, a single individual bears a number of different groups of organs which are fundamentally homologous through common origin (and not infrequently still connected by intermediate forms), but each of which has certain specific characteristics developed during its separate and independent (parallel) semophylesis.

It is, for practical purposes, not necessary to take LAM's paradox too much to heart: there are several degrees of homology, and there is nothing against the restriction of the search for semophyletic relations to sequences of forms separated, by levels of organisation, from older and from younger lines of descent, and against extension of the phylogenetic connections at either end when necessary. It is clear that the problem of the origin of the leaf must be studied in forms showing transitions from telomes to syntelomes, but the phylogeny of the various types of phyllomes need not be studied at the telomic level, the semophyleses of cataphylls, bracts, prophylls, sepals and petals being evolutionary processes involving leaves, not telomes. On the other hand, there is nothing to prevent us from relating, for instance, the genitalia of the Angiosperms with the sporangia of the Progymnosperms as long as we do not make sudden 'leaps' and derive stamens and carpels directly from ancient forms (as MELVILLE has done in his 'gonophyll' theory), thus skipping eventual important intermediate stages. Angiospermous ovules, though admittedly containing the derivative (homologue) of a mega-sporangium, possess integuments and sometimes arilli or other accessory organs. The unravelling of the evolution of the ovule must produce an acceptable explanation of the morphological nature of all its constituting parts and proceed from level to level, *i.e.*, from the sporangium to the ovule (with a single integument), subsequently from the unitegmic to the bitegmic condition, etc., until the whole sequence of events is re-constructed. I have previously already sounded a warning against those telomists such as WILSON and MELVILLE who immediately relate ramified systems of telomes and, *e.g.*, angiospermous pistils heedless of the intricacies of the evolution of these complicated organs.

If we return to the subject of homology, we can distinguish the follow-ing homology relations (compare Figure 2):

1. The direct homology of organs belonging to (and forming) a single semophylesis; there is one prototype and only one sequence of derived forms.
2. The indirect homology of organs belonging to different semophyleses in two or more parallel lines of descent based on a common ancestry (on a common basic taxon); there is one protoype but there are several parallel series of derived forms.
3. The indirect homology of organs belonging to different semophyleses in a single taxon phylogeny; there is also one prototype but there are several parallel sequences of derivatives.

This means that one can compare a given category of organs in a single semophylesis in one taxon phylogeny (according to the direct homology, case 1), or with other (similar) organs in other phylogenies, based on the same ancestral group, at the present-day level (according to indirect homology, case 2) or with other organs in different semophyleses belong-ing to the same phylogeny, also at the present-day level (according to case 3). Accordingly, one can compare homologies in *individual* semo-phyleses (in *lines*), and at any geological horizon, including the present-day (at coetaneous cross-sections or at the terminal ends) of *several* semophyleses (at *levels*). This is in order as long as we deal with or-gans which have a common prototype, because, from the scanty paleo-botanic records, we usually cannot reconstruct phylogenetic lines in such detail as to permit a more specific definition of homology than SIMPSON's (1961, p. 78): 'Homology is resemblance due to inheritance from a com-

mon ancestry'. The reconstruction of evolutionary lines being very much
a matter of opinion (or, rather, of intuition), a confusion of lines and of
levels belonging to a group of semophyleses based on a single prototype
with those belonging to different groups based on two or more prototypes is likely to occur. This is a constant source of diversity of opinions
and interpretations, as will be shown in the next chapter. The difficulty
is the frequent typological similarity or phenetic resemblances of organs
and characters in altogether independent evolutionary lines. Such similarities were probably best defined by SIMPSON (1961, pp. 78–79) for
taxonomic groups, but his circumscriptions are equally applicable to
morphological homologies and pseudo-homologies:

Parallelism is the development of similar characters separately in two or
more lineages of common ancestry and on the basis of, or chanelled by,
characteristics of that ancestry [*i.e.*, in fact the above-mentioned case 2, but
showing a similarity of features after a previous marked divergence].
   Convergence is the development of similar characters separately in two or
more lineages without a common ancestry pertinent to the similarity but involving adaptations to similar ecological status. Similarities so developed are
convergent.
   Analogy is functional similarity not related to community of ancestry. . . .
Convergent characters are analogous insofar as the similarity can be related to
function, which is usually and perhaps always the case.

These three notions are sometimes united as homoplasy, as opposed
to homology, but the difference between parallelism and certain forms
of homology is but a matter of degree, because parallelism is based on
an ancient propinquity of descent (see SIMPSON's definition of 'homology'!). Again I must emphasise that all these subtleties are worth but
little in practice, so that—in order to avoid circular reasoning—one
should primarily accept (postulate) the line of descent so as to have an
admittedly aprioristic but sound foundation. So far we have been dealing
with phylogenetic homologies, but there is another aspect of homology,
viz., the serial homology mentioned previously. Superficially, this relationship of organs based on similarity in development (in morphogenesis)
seems to agree with the above-mentioned case 3 of homology, because
it is tempting to identify the occurrence of typologically (phenetically)
different but homologous organs in a supra-specific taxon, often present
in a kind of sequential arrangement (even with transitional forms),
with the serial formation of different organs in more or less the same way
on a single individual. However, as stated before, the identity of serially
formed organs stands or falls with the identity of the morphogenetic
processes: if the latter are identical, very similar, and for all practical
purposes homologous, organs are consistently formed (*e.g.*, a sequence
of normal vegetative leaves), but if the morphogenetic processes are

appreciably changed (*e.g.*, as the result of a different induction), the organs formed before and after the change are not necessarily homologous and in either alternative there is no worth-while correlation between the phylogenetic homology and the so-called serial homology. In the first case the organs are identical, in the second presumably not even homologous, so that the serial homology has only a very limited application, its only phylogenetic aspect being the relation with the semophylesis of the shoot apex.

6
Evidence from Other Disciplines

Anatomy. Histogenesis and Organogenesis. Other Disciplines, Including Phylogeny.

Morphology started as a fairly isolated and self-contained branch of biological science based on scholastic principles (GOETHE 1790). To this day, a few morphologists tend to restrict 'phytomorphology' to an assembly of such conventional and Angiosperm-centred postulates, working hypotheses and deductions. LAM has recently (1962) compared the Old Morphology with a large and solidly built edifice provided with a single window permitting only the same limited outlook on the outside world. This is not such an exaggerated metaphor as one might think, because many a morphologist who flatters himself that he uses 'independent' ancillary evidence from other disciplines actually adapts these data to fit the preconceived morphological theory. Several examples of such practice are mentioned in other chapters (see, *e.g.*, Chapters 3 and 4). To make matters worse, one has been wont to forget that the idealistic morphology stood at the cradle of typological methods of classification, so that the supposedly corroborative results of traditional morphology and typological taxonomy are rather a foregone conclusion—being based on the same fundamentals, they jolly well have to be compatible! Such cases of perhaps unintentional but nonetheless deplorable circular reasoning which creates the false impression of a broad scientific approach can be found in many publications. Although the need for a broader outlook is almost universally advocated, it seems to be difficult to practise what is preached.

Anatomical studies are often similarly biased for the simple reason that before comparative anatomy can become interpretative it must be supplemented by a typological or phylogenetic evaluation of the respective taxonomic entities involved. An anatomical feature is usually declared to be 'primitive' or 'advanced' if it occurs in traditionally primitive or derived taxa, respectively, and inevitably some circular reasoning creeps in again. Nevertheless, exhaustive research has overcome this handicap up to a point. The long series of investigations by I. W. BAILEY (1954,

1956, 1957) and his associates have indubitably demonstrated that the 'woody Ranales' are a heterogeneous assembly because such rather fundamental differences were found between several subordinate taxa (differences which are, incidentally, to a large extent correlated with palynological characteristics) that not only is a close relationship between these groups virtually unacceptable but, also, the connections between other groups of the Angiosperms and the woody Polycarpicae become rather enigmatic. If one insists on the 'Ranalian' origin of all Angiosperms (or even only of the Dicotyledons), the anatomical indications are in fact so negative that no higher claim can be made than a 'remote' (*i.e.*, ancient) common ancestry, not to mention the moot point of indicating *which* subordinate Ranalian group has the closest propinquity of descent in respect of other angiospermous orders.

Phylogenetically speaking, the outcome of a great deal of painstaking anatomical research thus proves to be inconclusive and, in my opinion, emphasises the *isolated* position of such taxa as Magnoliales *s.s.*, Laurales, Nymphaeales, *Trochodendron* and *Nelumbo*. In several other chapters the relict status and the specialisation of the Polycarpicae generally (and the Magnoliales in particular) will be touched upon, so that for the time being the fairly obvious corollary will suffice, *viz.*, that the 'phylogenetic' (read: 'typological') derivation of non-Ranalian Angiosperms from forms with a magnoliaceous morphology is absolutely untenable. In other cases, too, the anatomical data are inadequate as indicators of the gross phylogenetic relationships. Rather paradoxically, the anatomical features have a considerable diagnostic value, mainly at the generic and family level. The importance of, *e.g.*, the structure of the secondary xylem for solving various taxonomic problems such as the placing of 'anomalous' or 'aberrant' genera, and the indication of the most likely alliance of certain variously classified groups (*e.g.*, Julianiaceae, near Anacardiaceae rather than Amentiflorae; Garryaceae, near Cornales-Umbelliflorae rather than Juglandales; etc.) need not be emphasised. I believe that there are several reasons why, upon the whole, anatomical indications have been rather disappointing as an aid in sorting out the phylogenetic relationships of larger angiospermous taxa. In the first place, the Angiosperms had already differentiated into a number of recognisable orders (even into some recent families and genera) by the middle of the Cretaceous, so that each discrete recent group represents the terminal members of a lineage that had led an individual existence of some duration, which implies an equally independent semophyletic evolution of its anatomical features. The probable polyphyletic origin of the Angiosperms was associated with an early divergent differentiation of their anatomical features at the still 'gymnospermous' level of evolution, and in this respect also they tended to drift farther apart during their subsequent Cretaceous and post-Cretaceous development. The likelihood of parallelisms also contributes to the complicated evolutionary history of the recent Flowering Plants. One of the axiomatic rules for instance, is that a certain type of vessel perforation (the scalariform end plate) is invariably more primitive than the large single perforation, but there is no reason to accept this as valid throughout. The Chlamy-dospermae provide examples of vessels derived from tracheids with bordered pitting; this could also be one of the semophyletic pathways leading to the functionally adapted xylem vessel in Angiosperms.

Another much discussed subject is the nodal anatomy (BAILEY 1956). After a period in which the trilacunar and three-trace leaf gap was supposed to represent the universal primitive condition among the Angiosperms, the unilacunar two-trace gap is more favoured nowadays. The fact is that in several, by general agreement more or less basic, groups not only three-, and two- and one-trace (treble, double or single) gaps occur, but also multilacunar nodes (e.g., Dilleniaceae, some Magnoliales; cf. OZENDA 1948), and that the occurrence of numerous traces is a rather typical 'cycadeoid' feature which is also characteristic of the Monocotyledons. A phylogenetic splitting of gaps (and of vascular strands!) to account for the presence of the multilacunar condition seems to me to be improbable in such primitive groups as Dilleniaceae and the early Monocots (e.g., Pandanales, see MEEUSE 1961a). Large-scale splitting (e.g., of stamens) has played a prominent part in older phytomorphological theories, but this is, to my mind, an incongruous and illogical assumption. Reductions, fusions and oligomerisations are of frequent occurrence, whereas 'splitting' goes completely against these normal phylogenetic trends. The varying number of leaf traces and nodal gaps demonstrates a number of ancient conditions, presumably going back to several already diversified early groups and, at the angiospermous level of evolution, already developed as *alternative* anatomical features, so that the nodal anatomy of the recent Angiosperms cannot possibly be a reliable yardstick to assess the *relative* advancement of this character in any given group. However, the coincidence of multiple traces and other primitive characters in Polycarpicae and Dilleniaceae indicates unequivocally that the multilacunar node more often represents a primitive condition than a secondary development. In this light, the transference of phylogenetic deductions based on the vegetative nodal anatomy to the interpretative morphology and anatomy of 'the' carpel (e.g., EAMES 1931, 1961) appears to be extremely speculative and has not the slightest demonstrative force, so that the assumption of a basic leaf-trace pattern (whether tri- or bi-lacunar, or simple with a double trace) common to all primitive stamens and carpels had better be discarded altogether. The argument also gets a different aspect if the genitalia are interpreted as fertile axes associated with subtending bracts (*e.g.*, MELVILLE 1960, 1962; see also Chapters 14 and 16). As regards other anatomical features (*e.g.*, the stomatal apparatus), embryology, and phytochemistry, one does not fare much better when attempting to establish unequivocally the relationships of the *major* taxa. In conjunction with the indications of the pollen morphology a postulation of ancient alternative conditions, of parallel lineages, seems to fit the observations best. The phylogenetic taxonomy based on recent forms thus remaining wide open to speculation, the phytomorphologist is deprived of the opportunity of basing his morphological deductions on an undisputed prototype.

The histogenesis or organogenesis (in the proper sense; usually and erroneously called the 'ontogeny') is another source of information. The ontogeny proper, *i.e.*, the development of a young individual from a seed, often provides some important indications by showing apparently ancient features, especially in the morphology of the first-formed leaves. The juvenile leaves may certainly be taken as an example of the 'law of recapitulation'—admittedly an unproved hypothesis based on intuition rather than 'facts'—and probably demonstrate the shape of the ancestral phyllomes. Of the numerous examples I have selected *Myrica javanica*, whose seedlings bear peculiar, deeply pinnatifid leaves with alternating lobes reminiscent of the fronds of some advanced (cycadopsid) Pteridosperms and Bennettitaleans, whereas the leaves of older plants have undissected serrate laminae (see Fig. 3). This case is also interesting in connection with the shape of the leaf in the American 'sweet fern' (*Myrica asplenifolia* or *Comptonia peregrina*) which is also deeply dissected and reflects the origin of the myricaceous (perhaps amentiflorous?) simple serrate leaf from a pteridospermous or cycadopsid alternipinnate frond. The ontogenesis does not provide such atavistic stages of the floral (reproductive) organs because the latter do not develop directly from the embryo but only after some vegetative growth has been formed and, unlike the leaves, never have 'juvenile' precursors. The developmental processes at the vegetative shoot apex do not constitute a true ontogenetic differentiation of a zygote into a whole or 'complete' organism, but rather the repetitive formation of organs in a regular sequence of plastochrons which arise, develop and mature, a process which can, in the vegetative phase, proceed more or less indefinitely, only to change when the apical meristem dies off or is transformed into a floral apex. In the generative phase of the shoot apex, different sets of organs are formed which, on the analogy of the vegetative phase, are identified as lateral (foliar, appendicular) or cauline (axial). Although this seems to be a straightforward method of interpretation, opinions are divided, and this can only mean that there are certain complications prohibiting the unequivocal (hence, unanimous) identification of a fertile organ with one of the morphological categories of sterile organs. A case in point is the independent histogenetic origin of the placentae and the sterile, encasing ovary wall in pistils with free central placentation. Some phytomorphologists (including myself) consider this to be a very strong indication of the phylogenetic independence of the placentae and the ovary wall, but other workers (*e.g.*, EAMES 1961, SATTLER 1962) maintain that such ovaries represent the extreme stages of a semophyletic sequence leading from a syncarpous septate ovary with 'axile' placentation to free central placentation through a progressive reduction of the

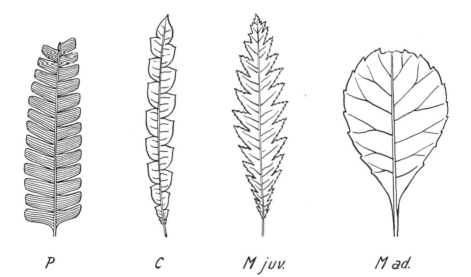

FIGURE 3. Atavistic juvenile stages, demonstrated by species of *Myricaceae*: *P*—An early cycadopsid frond. *C*—*Myrica asplenifolia* (*Comptonia peregrina*). *M juv.* = *Myrica javanica*, leaf of juvenile specimen (seedling). *M ad.* = *Myrica javanica*, leaf of adult plant.

septa, so that each placenta and the opposite portion of the ovary wall are derivatives of a single 'carpel'. A third explanation is based on the ancillary hypothesis of the peltate nature of the carpel. Even when the very same species were histogenetically studied (*e.g.*, some Caryophyllaceae and Primulaceae), the relevant publications report often diametrically opposed conclusions (see, for instance, the papers by ROTH 1959, PANKOW 1959, SATTLER 1962 and MOELIONO 1959 on the subject). The development of a number of so-called pseudo-monomerous ovaries was investigated by ECKARDT (1937, 1957), who maintains that they are carpellary with leaf-borne solitary ovules, whereas BARNARD (1957a, 1958), who studied some Gramineae and Cyperaceae, and PANKOW (1962) who examined a variety of forms, decided upon the cauline origin of the ovule ('stachyospory'). Protagonists of the appendicular floral theory (*e.g.*, TAKHTAJAN 1959a, b, EAMES 1961) claim that the ovules are 'leaf-borne', but SAUNDERS's (1937–39) theory of 'carpel polymorphism' is essentially based on the conviction that the placentae are a different kind of structure from the valvular portions of the gynoecium. The dual origin of a pistil from a fertile (and axial) placental organ and a sterile, associated bract-like element or from several such units is defended by an increasing number of botanists: HAGERUP (1934–38), FAGERLIND (1946, 1958), BARNARD (1957–60), MELVILLE (1960–63), MEEUSE (1963c). LAM (1948 *et seq.*) and PANKOW (1962) accept this origin for the ovaries of some but not of all Angiosperms. A few workers deny any correlation between fertile and sterile organs which implies a *sui generis* origin of the genitalia (GRÉGOIRE 1935, 1938, McLEAN THOMPSON 1934, 1937). The *sui generis* concept will be discussed in Chapter 7, where the independent origin of the floral appendages other than the sporangia is rejected.

Since the one morphogenetic or 'ontogenetic' school has not convinced the other, a kind of stalemate has been reached and there is some danger of a maintenance of the *status quo*. Histogenetic arguments have been adduced *pro* or *con* almost any given floral theory, which may create the impression that developmental studies cannot be the last word in morphological deductions. However, the accumulated evidence seems to favour the axial or cauline interpretation of the placentae and the stamens (see also Chapter 13).

Teratological data have been discussed in another chapter (see page 27), where the restricted value of abnormalities in interpretative morphology is emphasised.

Floral ecology has also been used as a source of phylogenetic speculation, but apart from several general conclusions, such as the impetus given to certain evolutionary processes by insect pollination and the likelihood of many parallelisms (zygomorphy, progressive reduction of the androecium and also the advent of biological flower types in unrelated groups, *e.g.*, 'trap flowers' in Araceae, Aristolochiaceae, *Ceropegia*, an occasional Orchidacea, etc.), such considerations do not contribute much in the way of morphological argument, chiefly because the workers in the field of floral biology almost invariably base their deductions on the Ranalian or euanthium theory of the origin of the Angiosperms. The study of genetics has been extremely useful in demonstrating possible mechanisms in evolution, but has very little bearing on the gross semophyletic and hologenetic processes.

The young scientific discipline called 'palynology' (the study of microspores and pollen grains) has contributed so many valuable data that phylogenetic botanists and particularly the classifiers would be foolish to ignore the pointers from palynological studies (ERDTMAN 1960a). Although the importance of pollen morphology in classification is perhaps still not sufficiently recognised, palynological characters tend to become an integral part of diagnostic taxonomy. Phytomorphology thus benefits indirectly in that the likely as well as improbable phylogenetic relationships deduced from pollen studies indicate those cases in which more or less direct semophyletic connection may be accepted and those in which there are contra-indications. The above-mentioned example of the Magnoliales is illustrative. the isolated position of this group with respect to the bulk of the Dicotyledons denying the customary derivation of the stamens and ovaries of all other Dicotyledons from magnoliaceous prototypes in conventional floral morphology. This does not mean that the palynologist always has the last word, because the difficulty of distinguishing between semophyleses, parallelisms and convergencies also cramps his style. I believe that the hitherto rather neglected study of the possible relations between the morphogenesis (*i.e.*, the microsporogenesis) and the ultimate morphology of the pollen grains may be very illuminating in this (and in other) respects.

7
The Beginning

Origin of Vascular Plants and of the Main Subdivisions.   Telomes and Telomists.   Application and Limitations of the Telome Theory.

The advent of the vascular plants is a matter of conjecture. The fact is that in the Lower Devonian strata the remains of several primitive but already somewhat differentiated types of terrestrial vascular plants have been discovered and these must, by that time, already have had a considerable geological history. This is substantiated by reports of still earlier (Silurian and older) vascular plants, but most of the more ancient remains are too fragmentary to enable a satisfactory reconstruction.

Various suggestions regarding the origin of the vascular plants or 'Tracheophyta' have been made, the most important contributions being those of BOWER, CHURCH and FRITSCH, but apart from phytochemical evidence, *viz.*, the correspondence in the types and relative quantities of the chlorophylls and of some accessory pigments, that they descended from the green algae or Chlorophyta, there is nothing tangible. It is rather generally assumed that the original habit of the sporophyte of the oldest transitional forms was a thallose one such as is found in several groups of Algae and in the frondose Liverworts,* *i.e.*, a more or less laminose growth form which was weak and must have been lying flat against the substratum (or floating in the water as CHURCH's [1919] hypothetical 'Thalassiophyta'), grew by means of apical (terminal) cells and branched dichotomously. There is only circumstantial evidence that such organisms, if they ever existed at all, were the ancestral forms of the Tracheophyta, and alternative hypotheses have been brought forward (see FRITSCH 1945); but a discussion on this point will be beyond the scope of this book. It is irrelevant in the following discussion, if we simply *postulate* thallose ancestors which had more or less branched green sporophytes and produced spores. Early in the phylogeny, the spores must already have been produced in distinct sporangia which

* In the gametophytic generation, though!

were situated near the periphery so that the spores could be released after maturation. This finds support in the observation that the oldest recognised Tracheophyta had peripheral or terminal sporangia. The palaeobotanic records also point to isospory, or at least to morphological isospory, as the most ancient condition, heterospory only appearing in younger groups. According to current views, the thallose unvascularised ancestors gave rise to vascularised forms and became 'telomic' (ZIMMERMANN). The thallose sporophyte changed into a (still dichotomously branched) system of more or less cylindric to ribbon-like axes, and in a considerable number of the recorded early vascular plants the sporangia were consistently apical on some or on all of the ultimate ramifications of the telome system. Such plants have been classified as Psilophyta or Psilopsida, but several workers (LECLERQ, AXELROD 1959, MERKER 1961, JEFFREY 1962) have recently expressed as their opinion that it is not at all certain whether the known fossil forms (and, eventually, the recent Psilotales or Psilotaceae) with this general habit indeed constitute a homogeneous and monophyletic ('natural') group and that the comparatively young Rhyniaceae (of Middle to Upper Devonian age) could not possibly be the *ancestors* of coetaneous and older, already more highly organised Cormophytes. It is, for instance, striking that the oldest Lycophyta, such as *Drepanophycus* and *Baragwanathia*, were plants with stem-like axes bearing helically arranged assimilatory organs ('Lyco-leaves') and *lateral* sporangia, so that it remains to be seen if they descended from plants of the habit of a *Rhynia*. This is also a warning against generalisations by telomists such as ZIMMERMANN who 'derive' all cormophytic groups from a telomic psilophytalean prototype virtually corresponding with the Rhyniaceae.

On the other hand, one can 'work backwards' and attempt to relate the more recent forms with older fossils by a process of extrapolation into the past. This method of reconstruction of the phylogenetic history has yielded the generally accepted 'over-all' lineages of some major taxa, such as the one leading from *Baragwanathia* via *Drepanophycus* and *Protolepidodendron* to the Lycopodiaceae, to the Selaginellales and (through the Lepidodendrales *s.l.*) to *Pleuromeia* and the Isoëtales, and, in the Sphenophyta, the series *Protohyenia, Hyenia, Calamophyton* and *Phyllotheca* to Sphenophyllales, Calamitales and Equisetales. A similar retrograde reconstruction of a lineage makes it highly probable that all gymnospermous groups and their recent descendants (including the Angiosperms) can ultimately be traced back to the Upper Devonian Progymnospermopsida *sensu* BECK (1960), and I am of the view that the extrapolation can be extended to such Middle Devonian fossils as *Protopteridium* and *Svalbardia,* which were practically 'telomic' forms with a 'psilophytalean' habit. This will provide sufficiently ancient prototypes for those groups, *viz.*, the Spermatophyta, that provide the principal subject matter of this book.

The remaining major group, comprising the true ferns, presumably goes back to similar and perhaps related prototypes (*e.g., Cladoxylon*) and most probably has gone its own separate ways since Middle to Upper Devonian eras. Thus we can retrace the individual main lineages of the Tracheophyta and arrive at a working hypothesis of the phylogenetic and semophyletic relationships in each of the major subdivisions of the Cormophyta which can serve as the starting point of a morphological analysis of each individual major group, commencing with the oldest and consequently primitive taxa, by following certain semophyletic trends in what appear to be hologenetic sequences of taxa leading to the geologically younger and, ultimately, recent forms. Phylogenetic relations and semophyleses have been elaborated along these lines and, as far as the Lycophyta (Lycopsida) and Arthrophyta (Sphenopsida) are concerned, the over-all results are fairly clear, rather unequivocal and apparently uncontested, so that it will suffice to refer to the relevant publications, *e.g.*, those by LAM and ZIMMERMANN. The true ferns need not be discussed at length either, because they are another, though probably rather heterogeneous, assembly of forms with a long geological history that have hardly changed since Palaeozoic to early Mesozoic times, provided only that we include the numerous (and presumably partly polyrheithric) isosporous groups and exclude the heterosporous water ferns, which I think are surviving Pteridosperms.

Although I recommend the separate analysis of the four main subdivisions of the Cormophyta, this does not mean that certain aspects of the semophyleses and morphogenetic processes in these major groups cannot be studied from a common point of view. If the retracing of the lineages into the past leads to telomic ancestors, we may certainly attempt to apply the general principles of the telome theory, as long as we do not confuse the general trends with the special evolutionary developments of individual lineages and keep in mind that, although we can speak of leaves, stems, roots, bracts, stegophylls, etc., in all major groups, it does not imply that these organs are invariably homologous throughout these groups and that, for instance, the assimilatory organs (the *functional* 'leaves') of Lycophyta, true ferns, Conifers, Pteridosperms and Flowering Plants are morphological equivalents. It would be too cumbersome a task to coin names for all the analogous and functionally comparable organ categories in the different groups (some have already been made, such as 'Lyco-leaves', 'enation leaves', micro- and megaphylls, etc.) and we need not drop the general terminology (stem, root, bract, etc.) as long as we restrict the morphological and phylogenetic implications of 'resemblances'—the basis of our semophyleses—to a single monophyletic group only. Similarly, we must not morphologically relate parallel developments in the main subdivisions such as heterospory and the occurrence of secondary growth, although we may accept the rather vague notion that identical morphogenetic evolutionary forces 'of selective value' may have caused these 'convergencies', parallelisms or analogies. In practice it is not at all easy to distinguish between homologies and pseudo-homologies, and one must constantly be on the alert to avoid these morphological pitfalls. This subject is deemed important enough to devote a special chapter to it (see Chapter 7).

The telome theory has been exposed in several publications by ZIMMERMANN, LAM, WILSON and others, the latest comprehensive treatment being that in ZIMMERMANN's textbook of 1965. There are a few fundamental principles on which I do not see eye to eye with ZIMMERMANN and other telomists. In the first place, the sporangium is supposed to be a part of a fertile telome and such a fertile telome is regarded as fundamentally homologous with a sterile telome. The production of spores is a very old feature of primary survival value already occurring among the Thallophyta and must have preceded the differentiation of a thallose frond into telomes, so that the sporangia most probably existed before the Protocormophytes attained the telomic stage (or any other possible early Tracheophytic growth form, for that matter) and may be regarded as organs sui generis (see also MEEUSE 1963b). The sporangium is older than the leaf (BOWER) and this fundamental conclusion is important in connection with the 'sporophyll' concept, as we shall see in another chapter. One may quibble about the nature of the sporangium wall and consider it to be a part of the supporting telome (sporangiophore), restricting the qualification 'sui generis' to the sporogenous tissue, but this is only of academic interest. The sporangium is a morphological entity, so that it is more convenient and logical to regard the whole sporangium as sui generis in our discussions. For practical purposes, the meaning of the definition of organs sui generis is that, when fertile, they have no homologous sterile counterpart in the same individual, or vice versa. In the case of the sporangia and their homologues, such as the nucelli of the ovules and the locules of the anthers, it renders the question whether an ovule is an axis, a bud or a leaf (a much-debated issue in the years between 1860 and 1900 and as recently as 1956 again touched upon by MAJUMDAR) altogether futile—there simply are no truly sterile counterparts and sporangia constitute a separate category of organs. Only the sterile portion of what ZIMMERMANN calls a fertile telome, which can be referred to as a sporangiophore, is a derivative of a sterile portion of the pre-telomic thallose frond and consequently completely homologous with a sterile telome. The elementary morphogenetic processes distinguished by ZIMMERMANN in his telome theory, such as planation, fusion (webbing), incurvation and overtopping, apply to both sporangiophores and sterile telomes, and also to combinations of the two. The sporangia did not 'actively' participate in the first stages of the semophyleses, but shifts in their relative positions were of course brought about by the phylogenetic changes involving their sporangiophores. This does not seem to be so essential, but, as we shall see later on, it is important in connection with *sui generis* theories concerning organs of the fertile regions of the Higher Cormophyta.

Another consequence of the *sui generis* nature of the sporangia concerns the interpretation of those categories of sterile organs which are evidently the homologues of sporangia or ovules and can be defined as sporangia (ovules) that have become sterile through loss of their sporogenic tissue. Again, for practical purposes, it is best not to enter into hair-splitting discussions about the interpretation of the residual constituents of such sterile organs, but—the fundamental homologies being clear—to refer to such organs as 'sterilised sporangia' or 'sterile homologues of sporangia (or ovules)'. This is seemingly in contradiction with the previous statement that organs *sui generis* have no sterile homologous counterparts, but this is not the case if we consider the homologies of (phylogenetically speaking) originally sterile organs (sterile telomes and their derivatives) as opposed to those of phylogenetically fertile organs (sporangia and their sterile derivatives), *i.e.*, as belonging to two completely different categories which have had their own phylogenetic history since the pre-telomic level of evolution and are, properly speaking, both *sui generis* in respect of the other.

That is why I think that the fundamental distinction made in the telome theory between fertile and sterile telomes (and mesomes, which are only a special, *viz.*, a non-terminal, kind of sterile element) is not at all satisfactory and that one should distinguish (1) fundamentally sterile elements derived from and homologous with the sterile parts of an undifferentiated thallose sporophyte, including telomes (mesomes) and sporangiophores, and (2) fundamentally fertile (sporogenic) elements, comprising the sporangia and their sterile derivatives.

Another objection I have against some declared telomists is the overrating of the applicability of their theory to the most specialised Cormophyta. When LAM (1959a) contends that even the Angiosperms are, as it were, still 'full of tiny *Rhynias*', his statement, apart from being an unintentional false metaphor—the sporophyte of an Angiosperm can only be the homologue of a single *Rhynia* sporophyte!—is dangerously suggestive of a simple and 'ultimate' explanation of the morphology of the Angiosperms. This is an unwarrantable simplification. Not only does the principle that an entity is more than the mere sum of its parts apply, so that among the Higher Cormophyta various interactions and correla-

tions between the organs developed, but the individual organ categories also underwent semophyletic changes which no longer reflected
the fundamental telomic origin. *The only character in which all Higher
Cormophyta still agree with their psilophytalean ancestors is the fundamentally terminal position of the sporangia.* In addition, some Higher
Cormophyta show more or less distinct traces of the ancient dichotomy
in *some* of their organs, but it is not so easy to weed out all those
secondary and false dichotomies that are not remnants of the ancient
branching system. VAN DER HAMMEN's (1948) long list of dichotomies
contains some presumably genuine examples, such as the branched stems
of some Pandanaceae, palms, *Dracaena* and *Aloë*, and some types of
stamens (to which the venation of the leaves of *Kingdonia* and *Nelumbo*
may be added), but it is an exaggeration to see an ancient dichotomy
in all bifurcations of veins in Angiosperm leaves, in the stigmas of
Begonia or in the forked leaves of *Drosera binata*. The danger lies in
the fact that *every* bifurcation of vascular strands may be identified with
a dichotomy of telomes and mesomes (and *every* strand with the central
stele of a telome or a mesome), the whole phylogenetic history of the
organ, the possibility of physiological adaptions and other selective
evolutionary forces being disregarded. This is especially hazardous if
such phenomena as serial splitting and axillary relationships, *i.e.*, associations of fertile and their subtending (axillant) sterile organs, are simply
'explained' as ancient dichotomies (see, *e.g.*, LAM 1961b). MELVILLE
(1960, 1962, 1963) derives his 'gonophylls' from ancient dichotomously
branched telome systems of which one half was sterile and the other
fertile, the sterile part of a female gonophyll ultimately providing the
sterile portion of the Angiosperm 'carpel' and the fertile part the placental
portion (for details see Chapter 13). His suggested derivation completely ignores the fact that an ovule is not a sporangium but a sporangium homologue enveloped by integuments, sometimes also by an arillus,
i.e., an interpretation of these integuments; the arillus and other ovular
or placental organs must *precede* the interpretation of 'carpels' and
'ovaries'. In some of the following chapters such theories will be
critically discussed in more detail, and altogether different alternative
explanations will be suggested.

A more realistic approach is to restrict the application of the telome
theory to telomic and syntelomic organs in geologically old groups and
to the early evolution of 'leaves', 'stems', 'fronds', etc., out of branched
telome systems. Once a syntelome assumed the morphological characteristics of an individual organ (such as a leaf), the whole organ no
longer behaved as a loose assemblage of telomes but as an entity. The
homologies are rendered even more intricate by the biochemically controlled correlations and other mutual relationships between 'lateral' and

'axial' organs and by the interaction of morphogenetic processes and inherited morphological structures, to be discussed in a special chapter devoted to the Angiosperm 'carpel'. This interaction is partly attributable to the hitherto rather neglected semophylesis of the shoot apex. The traditional and originally Angiosperm-centred distinction between 'lateral' and 'axial' organs ('leaves' and 'stems'), a distinction carried through to the reproductive organs, rests in the Higher Cormophyta ontogenetically (morphogenetically) on the different origin of these two categories from different parts of the shoot apex. In principle, axial organs are the direct derivatives of the central core of the shoot apex whereas the 'lateral' organs develop from lateral appendages of the apex, the primordia. If one compares the origin of a telome dichotomy, which can only be the result of an equal division of a telome apex, with the complicated conditions in the apical region of the shoots in the Higher Cormophyta with a 'sympodial' construction, it is obvious that the explanation of the differentiation of a branched telome system into a stem and its appendicular leaves is not simply the result of overtopping, planation and webbing, but must have involved the shoot apex; and this again must have been attended with evolutionary changes in the morphogenesis, *i.e.*, in the subtle biochemical control mechanism of the areas of growth and differentiation. An additional complication is the change, quite certainly also biochemically (phytohormonally) induced, from a vegetative phase of the shoot apex into a fertile stage when the Higher Cormophyta start 'flowering'. This is manifestly also the result of a long phylogenetic history of the apical zone and the concomitant evolution of the biochemical control mechanism of the processes of rhythmic growth and periodic differentiation in meristematic tissues. A comparison of such involved and interrelated situations with the simple alternative of a 'fertile telome' and a 'sterile telome' in psilophytalean prototypes should put us on guard against simplifications. I shall be the first to admit that for a morphological analysis of semophyleses the telome theory provides a comprehensive and useful interpretation of the initial phase or phases, but in more highly organised forms we must reason in terms of the new syntelomic complexes that have become individual and more or less autonomous organs (stems, leaves, synangial complexes of sporangia, etc.) After all, dynamic morphology is not simply the chasing of telomes, but a hunt for semophyleses, for *organ* lineages.

8
'Lines' and 'Levels'

Definitions.   Confusion of 'Lines' and 'Levels' and Its Implications.
Some Advances Resulting from Re-appraisals of 'Lines' and 'Levels.'

As far as can be ascertained, LAM (1954) was the first to
speak of the confusion of 'lines' and 'levels'. By coining these terms, he
did not introduce new concepts. The subject matter had been discussed
before, for instance, in terms of (phylogenetic) homology and analogy,
but LAM's rendering is so concise and lucid that relevant cases can be
much more summarily discussed (and, incidentally, it also supplies a
convenient heading for this chapter).

By a 'line' is meant a line of descent, a phylogenetic lineage, also
referred to as a genealogy or a genorheithrum, that is, the series of
consecutive generations directly connecting an ancestral group and any
taxon descended from it. However, as a rule, it is not the phylogeny
that is intended but only one of its semophyleses, in other words the
continuous sequence of stages in the evolutionary history of an organ or
organ system. The essence is that the successive stages of the organ
or organ system are directly homologous. For 'organ' we can of course
also read 'character' in most cases, or 'taxon' if we wish to refer to a
complete phylogeny.

The 'level' refers to a corresponding degree of development of a
given organ in more than one evolutionary lineage and is an abbreviated
descriptive term for the morphological similarity of a stage of evolution
of two or more organs (or correspondence of typological characters)
attained in different 'lines'. To preclude any misunderstanding, a 'level'
is not a stratigraphic level nor does it imply that the resemblance of the
organs (or characters) under discussion are or were necessarily reached
simultaneously, *i.e.*, at the same geological time level. The 'level' of
differentiation is simply a collective circumscription of cases of homoplasy (analogy, parallelism or convergence).

Phylogenetic and taxonomic relationships are based on the concept
of homology, *i.e.*, on propinquity of origin (descent from a common
ancestral taxon), so that any assumed homology which is in fact an

analogy may result in a grouping of heterogeneous, phylogenetically unrelated elements into a single genealogy, and *vice versa*. This is the
confusion of 'lines' and 'levels', in LAM's terminology, which should and,
in theory, can be avoided, but in actual practice remains very much a
matter of personal opinion, simply because, inescapably, one follows a
circular reasoning in defining a phylogenetic relationship, a homology
and an analogy, a 'line' or a 'level'. All these notions are interdependent,
because they are not based on different postulates. The fundamental assumption is that there is organic evolution, hence phylogenetic relationship through common origin (propinquity of descent), and from this
starting point the concepts of true phylogenetic homology, *i.e.*, genealogies and semophyleses or 'lines', and their opposite numbers (analogies,
'levels', etc.) are subsequently deduced. However, one has to begin
somewhere and, as far as I can see, the only way is to try to establish
a kind of statistical probability that certain taxa are related. This
supplies a phylogeny or a number of phylogenetic lineages and the rest
follows logically. Experimental taxonomy and genetics, as well as some
other disciplines mainly dealing with taxa of lower rank (at the generic
level and below) strengthen our (let us admit it) intuitive conviction
that the higher the number of resemblances, the closer the relationship.
The resemblances (including homologies and analogies!) are often
identified with 'correspondence of characters', but this is fallacious, and
so is the confidence expressed by some authors in their ability to distinguish the true corresponding characters (the homologies) from the
false ones (the analogies). Admittedly, taxonomic experience counts,
but it also preaches caution, and most taxonomists are well aware of
the fact that other, much more complicated 'resemblances', such as a
correlated correspondence of a number of phenetic similarities, may reflect a convergence of two or more independent groups of correlated
characters. SIMPSON (1961) mentions several examples of convergence
which led to a very high degree of similarity, such as the resemblance
between the Tasmanian wolf (*Thylacinus*, a marsupial) and the true
wolves (carnivores, Canidae), between the extinct one-toed South
American ungulate *Thoatherium* (Litopterna) and the horses (Equidae),
and the often-cited convergence of streamlined body, shape of the extremities, position of nostrils and eyes, etc., of aquatic and secondarily
aquatic vertebrates.

A botanical analogy is found in the succulents, comprising unrelated
taxa showing a most striking convergence in habit and other characters
(such as loss or extreme reduction of leaves in stem succulents, perhaps
also the singular physiological and biochemical adaptation known from
a few species, *i.e.*, opening of stomata and fixation of carbon dioxide as
an organic acid during darkness, and photosynthesis with closed stomata

during the day). A correspondence of sets of such interrelated characters, all connected through a single 'adaptive' evolutionary trend, is hardly of more significance than a convergence of a single character; and who would claim to be infallible in distinguishing independent characters from correlated ones?

As SIMPSON puts it, we *must* accept the view that a close resemblance and a good correlation of characters which we can 'observe' or 'establish', coupled with the lack of any contra-indications against their homology, is indicative of a binding principle, a common source of 'correspondence'. I agree that in many cases this typological relation reflects the phylogenetic relation, but it is much more to the point if one simply states that a certain phylogenetic relation is *postulated,* so that all observations and other available evidence can be confronted with the postulated relationship and with the deductions following from this same apriorism. I add the warning that the judgment of the correspondence of characters is subjective, but we may feel more confident if the correspondence is demonstrable not only by typological agreement of 'morphological' or 'phenetic' characters, but also by a consistent correlation with genetical, cytological, physiological, ecological, phyto- and biochemical, chorological and other data (even including such corroborative evidence as host-parasite relations and the behaviour of animals). The *degree* of correspondence, *i.e.*, the relative relationship, remains a matter of opinion, because one worker tends to consider some characters to be of more 'importance', *i.e.*, as having more demonstrative force, than others, whereas another taxonomist or morphologist selects a different set of such 'major' (or 'key') characters and arrives at a different arrangement.

It is perhaps not generally realised that one can hope to convince a majority and achieve an adoption of one's suggested taxonomic arrangements, putative phylogenetic relationships and probable organ semophyleses only by a consensus of opinion. Very distinct (also called 'isolated', 'homogeneous' or even 'natural') taxa are usually accepted by a great majority (the living Cycads and several Angiosperm families such as, *e.g.*, Compositae, are a few of many examples), which is a logical consequence of our way of reasoning: there is a high degree of correspondence among all members of the group and a considerable lack of resemblance (a 'gap') between the members of this taxon and those of all other taxa, so that most authorities cannot help reaching the same conclusion, *which becomes the 'established' (standard) opinion by majority rule.* There is, however, a great deal of disagreement even as regards groups considered by a large majority to be 'homogeneous'. The two examples I have selected are illustrative. The position of the undoubtedly 'aberrant' genus *Cycas* in respect of all other recent genera of the Cycads can be regarded a sufficient reason to divide this taxon into two taxa of equal rank, one comprising the genus *Cycas* and the other the remainder of the genera, which is synonymous with the assumption of a least two parallel evolutionary lines that must have led an independent existence of some duration. The Compositae are 'split' by some authorities into two families, *viz.*, the Cichoriaceae (=Liguliflorae) and the Asteraceae (*s.s.*), respectively, and, in my opinion, indeed a convergence of two or more lineages towards the 'type' of the recent Compositae is not altogether inconceivable.

As far as practical—or applied—taxonomy (herbarium and museum systematics) is concerned, such changing and re-arranging of taxa may have a considerable nuisance value in that it leads to inconsistencies, *e.g.*, in the filing systems adopted in different herbaria and in the sequence, the delimitation and the classification of taxa followed in different manuals and floral works, which require adjustment by practical taxonomists, lecturers and students whenever they have to 'change over', from a familiar 'system' they have grown up with, to a different arrangement. The inconvenience causes them to dislike and even distrust any suggested major regrouping, simply because the 'old system' worked well in practice and the 'new arrangement' is not necessarily an improvement. There is of course no *essential* difference in the practical application of any consistent 'system' as long as every taxon can be referred to its own cubby-hole, properly tagged with a label, conveniently assembled in categories and satisfactorily keyed out.

Speaking as a practical herbarium botanist, I am certainly against any immediate drastic re-arrangement, for exactly the same reasons—it would serve no useful purpose, because, considering that, at least in theory, many taxonomists strive at a 'natural', *i.e.*, phylogenetic, system of classification, it would be decidedly premature to propose important alterations in the classification of the Higher Cormophyta until there is sufficient evidence to produce a first approximation of the true phylogenetic relationships that will be acceptable to a large majority of taxonomists and promises some degree of finality. As examples of established classifications, the arrangement of the true ferns and of the mammals can be mentioned; it is significant that in either example fossil evidence and typological methods have been aligned, so that the phylogenies of the major taxa have become fixed or almost so and the mammalogists and pteridologists have achieved a satisfactory degree of stability and 'uniformity of treatment'. On the other hand, exerting prudence to avoid premature changes should not lead to conservatism and procrastination. We can follow the example of mycologists such as my countryman, Dr. M. A. DONK, who have decided that the time is ripe to commence rebuilding the classification of the Fungi, starting from the very foundations, and do not hesitate to propose the most radical changes. Some day we shall have to tackle the far less laborious task of 'cutting up' and reclassifying the Angiosperms, but we must only attempt it when stability and uniformity are within reach.

In phylogenetic botany, unlike herbarium systematics, the re-arrangement of taxa often has much more far-reaching implications than a mere inconvenience. As a result of the splitting of a taxon, a previously assumed single phylogeny becomes a pair or a multiple of evolutionary lines, each with their individual semophyleses, so that some previously assumed homologies become analogies (or other homoplasies). Conversely, every 'lumping' leads to a recognition of a homology instead of an analogy; 'lines' become changed and 'levels' previously supposed to be convergencies or parallel developments become stages (sequential semophyletic phases) of a single 'line'. The various notions depend in the first place on the postulated taxonomic arrangement, and if a worker accuses a fellow-botanist of being a 'linist' or a 'levelist' it is often a case of the pot calling the kettle black. However, if only a *single* character is considered to be decisive or highly significant, especially if existing and fairly acceptable classifications are disregarded, one can easily be caught in the pitfall of making 'lines' out of 'levels'. A characteristic example is EMBERGER's (1949, 1952) distinction between 'Préphanérogames' and 'Phanérogames', which cuts through established major taxa (the main '*lines*') merely because his single criterion rests upon the degree of evolution of the 'seed' (which is nothing but the ultimate phylogenetic '*level*' of the retention of the megasporangium and its derivatives on the sporophyte). Lepidocarpales, Coniferales, Cycadeoideales (Bennettitales) and Angiosperms all have a more or less 'complete' stage of the seed, but they belong to at least three independent 'lines', whereas several major lines of descent, expressed in the fairly generally accepted vertical phylogenetic relations between Cordaitales, Ginkgoales and Coniferales, and between Cycadales, Cycadeoidales and Angiosperms, each contain a prephanerogamic and a phanerogamic 'level'. EMBERGER's grouping would also result in the inclusion of the genus *Gnetum* (Chlamydospermae) and even of some Chloranthaceae (Angiosperms!) in the 'Préphanérogames', because VASIL and YOSHIDA (1959) have shown that in *Gnetum* and in Chloranthaceae, respectively, the seeds, when shed, contain only a minute embryo that continues to develop after shedding, a 'level' that is found, *e.g.*, in *Ginkgo* and the Cycadales. Nobody, presumably not even EMBERGER, would agree to to so drastic a dismemberment as removal of the Chloranthaceae from the Angiosperms. It would raise the question of what to do with the remainder of the Piperales, undoubtedly related to the Chloranthaceae (in the same 'line'), and this would set off a chain reaction of absurdities.

Another example is FLORIN's well-known distinction between haplocheilic and syndetocheilic stomata, which is one of the key criteria used in referring fossil leaves (*i.e.*, also form genera) to 'Cycads' or to 'Cycadeoids'. It is one of the chief arguments used in discussions about the relationships of these two groups, not only to support the supposition (in my opinion, well-founded) that Cycads and Cycadeoids are not closely related, but also to classify fossil remains in the one category or the other. However, if one compares SAHNI's account of the stomata in the Pentoxylales with VISHNU-MITTRE's (1953, 1957) subsequent treatment of the same subject, it is quite clear that this was not a clear-cut case and also that the Pentoxylales, by general agreement of Bennettitalean affinities, do not conform to FLORIN's classification. MAHESHWARI and VASIL (1961a, b), after studying the development of the stomatal apparatus in *Gnetum,* concluded that the mature stage is variable and does not necessarily reflect its development (from a single mother cell or from several, the basis of FLORIN's definitions of 'haplo-' and 'syndetocheilic', respectively). Especially this last argument weakens the rigid and seemingly fundamental classification of fossil material (without juvenile stages!) into two presumed clear-cut 'lines', a haplocheilous line and a syndetocheilous line, a classification which cuts through the Chlamydospermae and other groups of Bennettitalean affinities as well. The morphology of the stomatal apparatus is, to my mind, not such an infallible criterion and certainly cannot prevail as a key character over other characters to indicate major 'lines', so that FLORIN's classification does not stand in the way of the existence of Bennettitalean (*s.l.*) groups with either type of stomata in mature leaves, or of the assumption that the Chlamydospermae belong to a single main 'line' and have Bennettitalean affinities. The devaluation of the stomatal character rids us of too static a framework of 'lines' and leaves much more scope for the appraisal of other characters and considerations, an appraisal which, in this case, results in the recognition of more Bennettitalean 'lines' instead of a single one and of a single phylogeny for the Chlamydosperms instead of two, whilst it also permits a closer propinquity of the Chlamydospermous and Bennettitalean lineages. In my opinion, these alternatives dovetail much more satisfactorily with other evidence,

The tenacious adherence to a certain interpretation of a morphological feature of a single taxon which is raised to the importance of a key character has also hampered the recognition of 'lines', of which I shall discuss here only one of the worst cases. That one is the well-known example of *Pleuromeia*, which was originally described as having its sporangia abaxially attached to the sporuliferous stegophylls. British palaeobotanists such as SCOTT and SEWARD pointed out long ago that, in view of this unusual situation (all other Lycophyta are known to have the sporangia adaxially attached), the supposed position of the sporangia was simply due to an error of interpretation and that, as the result of pressure on the stegophyll of *Pleuromeia* before or during the initial stage of fossilisation, the adaxial sporangium was squeezed through its supporting stegophyll and bulged out on the opposite side, so that it is nothing but an artefact. This process is familiar to everybody who has seen the effect of pressing on herbarium specimens of *Isoëtes* and is a result of the size, the shape and the solid consistency of the sporangium, which takes the pressure before the thinner stegophyll does and acts as a punch, so that it gets driven through the softer tissues of the stegophyll and bulges out on the opposite (abaxial) side. However, HIRMER maintained that the original interpretation was correct and in recent textbooks nearly all authorities accept his views, although some do not fail to point out the inconsistency; see, *e.g.*, ZIMMERMANN's (1959) discussion of the topic. HIRMER's interpretation has the following consequences: (1) *Pleuromeia* is a completely isolated taxon of which neither ancestral forms nor descendants with abaxial sporangia are known; (2) there is a 'gap' in the geological sequence of the Lepidodendrales and their putative descendants, the Cretaceous *Nathorstiana* and the recent genera *Isoëtes* and *Stylites;* and (3) although in all other morphological features *Pleuromeia* agrees with either the older Lepidodrendrales or the recent Isoëtales (or with both), it cannot be in the same 'line' with them—all this on account of the single prohibitive and paramount key character of sporangium attachment. If we drop HIRMER's interpretation and ascribe the aberrant position of the sporangium to a shift as a result of pressure before or during fossilisation, the following alternative theory can be defended: (1) *Pleuromeia* is morphologically related to older Lycophyta and recent forms and in fact intermediate in a number of characters such as size, adaptation to an aquatic habitat (*Pleuromeia* most probably grew in 'groves' at 'oases' in an arid area in very much the same way as date palms in the Sahara Desert), and development of the basal swelling of the stem and its lateral appendages (the latter presumably homologous with stigmariae); (2) there is a sequence in the occurrence of the palaeophytic Lepidodendrales, the mesophytic *Pleuromeia* and the recent Isoëtales; and (3) its morphological features connect *Pleuromeia* with the Lepidodendrales and the Isoëtales, so that it indeed fulfils all the requirements for a connecting link in a line from the former to the latter. In view of so much corroborative evidence, I believe it is high time that in forthcoming manuals *Pleuromeia* be assigned its logical and proper place in the classification of the Lycophyta.

Several palynologists (*e.g.*, HUGHES 1961a) deny the occurrence of 'angiospermous' pollen before the late Jurassic or early Cretaceous, which implies that 'the Angiosperms' as a group cannot be much older. This contention is based on the appraisal of the *level* of a single character, gymnospermous pollen grains being assumed to be inaperturate and angiospermous pollen grains provided with colpi or pori. However, there are anacolpate (anaporate) Angiosperms (*e.g.*, some Piperales and Aristolochiaceae); and how can one visualise the semophyletic development of angiospermous pollen grains from their gymnospermous proto-types without accepting intermediate forms such as Jurassic angio-spermoid sporomorphs? The 'level' of angiospermy is a combination of phylogenetic 'levels' of characters, such as anatomical features (*e.g.*, the formation of wood vessels), protection of the ovules (angiovuly), double fertilisation (and endosperm development) and the formation of a 'complete' embryo (a seed with a dormant phase). Each character has developed more or less independently and in some Angiosperms the ultimate level of all characters is not attained (*e.g.*, vessel-less secondary xylem and an 'incomplete' seed in *Sarcandra,* primitive stages of angiovuly in, *e.g., Canacomyrica* and the juglandaceous *Engelhardia*). The aperturate pollen grain is another Angiosperm character and must have developed (presumably polyrheithrically) from the inaperturate condition. The absence of typical angiospermous pollen types in early Mesozoic strata does not (as I see it) mean that early Angiosperms did not exist in those geological periods. An Angiosperm cannot be defined on the strength of a single character, on the basis of one 'level'.

There are indications that more 'lines' and hence more 'levels' must be recognised than is generally assumed, in other words, that there is much more polyphyly (polyrheithry), convergence and parallelism than is apparent from a perusal of the literature. Particularly, the development of such phenomena as complete siphonogamy (with or without double fertilisation) and 'angiospermy', and of such morphological structures as embryo sacs, functional 'flowers' (and 'flower types' of floral biology), fruits and seeds must have occurred in several independent or at least parallel lines. The monorheithry of several major taxa has been grossly exaggerated (*e.g.*, the supposed monorheithric origin of the Angiosperms, which is ultimately based only on the presumedly exclusive and singular character of double fertilisation; see MEEUSE 1964a). LAM divides the Angiosperms into a stachyosporous and a phyllosporous group (with an eventual intermediate group) as a reflection of two basic evolutionary lineages, having a stachyosporous and a phyllosporous ancestral group, respectively. I am of the opinion that the *level* of pseudo-phyllospory was reached in a number of individual, all initially stachyosporous phylogenetic genealogies (lines) and this leads to a completely different phylogenetic classification. This is another illustration of the far-reaching implications of the different interpretation of 'lines' and 'levels', in the case under discussion involving the origin, the evolution and the present-day taxonomy of a very large group of plants. Additional phylogenetic lineages become recognised when previously fragmentary records of a fossil group are 'linked up', so that its vertical and horizontal relationships can be assessed. The late Devonian Progymnospermopsida, to be discussed presently (as well as in several other chapters), provide a good example. The vertical relationships with the various groups of the Coniferophytina and the horizontal relationships with the early seed ferns complete the picture of the main phylogenetic lineages in the early Spermatophyta. The recognition of the Pentoxylales as an unknown major taxon, after the painstaking researches of Indian palaeobotanists, posed a problem because phylogenetic botany was saddled with an isolated group, of unknown alliance, but its Bennettitalean affinities were fairly soon established; in my opinion, the Pentoxylales are at the base of a previously unknown single 'line' leading to some, or all, Monocotyledonous orders (MEEUSE 1961a).

As we have seen, both in the last example and previously, classification comes first, whether we like it or not; and usually, at any rate in fossil material, this is based mainly if not exclusively on typology. However, additional and newly acquired evidence from palaeobotanic discoveries and morphological considerations, from postulated semophyleses and other sources, can and must be used continuously as checking material for the lesser or greater probability of the whole mental picture of phylogenetic relationships, 'lines', semophyleses and classifications. If there is good agreement between new observations and the postulated basic lines of descent, that is, if the additional data dovetail satisfactorily with the previously recorded or postulated subject matter and corroborate our deductions, we may feel confident without becoming overly optimistic, but if there are some clashes and cogent inconsistencies we must be prepared to abandon a traditional or cherished system of classification and its implications. These implications include the *morphological* classification, *i.e.*, the grouping of typologically homologous organs into the same category as against, in terms of phylogenetic botany, the grouping of all organs of common origin into the same semophylesis. Here another complication, which has actually obscured the recognition of semophyleses, arises, when postulated (typological) organ categories cannot be linked up with prototypes known from fossils. Especially that cumbersome legacy from the idealistic morphology, the 'sporophyll' concept, has been the source of much confusion and misconception and has been the sole reason why the origin of the Angiosperms remained a baffling mystery for such a long time (for about a century after the publication of DARWIN's *Origin of Species!*). The confusion of 'lines' and 'levels' was twofold in this particular instance: various phylogenetically non-homologous sporangium-bearing organs were classified as 'sporophylls' (the 'sporophyll' cult made botanists search for a 'sporophyll' that had to be found at all cost even though it was simply not there, which led to such absurdities as identification of integuments and Bennettitalean interovular scales with 'sporophylls'); and the hypothetical ancestral group of the Angiosperms was supposed to have had 'sporophylls' (or 'carpels') of a certain preconceived type. Both contentions are at least partly erroneous: the first because one and the same postulated morphological category manifestly contains analogues from several major evolutionary lines; and the second because although some ancestral protangiospermous and primitive angiospermous taxa can be indicated with a reasonable degree of confidence, they do not possess such 'sporophylls'. The consequences of such fundamental re-appraisals are far-reaching, and that is why I think it fitting to conclude this chapter with a few more examples of the profound and often gratifyingly elucidating effect of changing views regarding 'lines' and 'levels'.

There is a rather general consensus of opinion that the true ferns are an early 'offshoot' of the Cormophyta and have evolved independently (since Devonian eras, I believe). This means that there is no direct homology between the characteristic fern features (such as the indusium and the annulus) and certain organs of the other (higher) Cormophyta (to choose but a few examples, between a sporuliferous fern frond and a cycadopsid so-called 'sporophyll', or between an indusium and an integument); any resemblance is caused by analogy (convergence). Personally, I am inclined to assume that, at any rate since the Carboniferous, the 'level' of heterospory was never attained by the ferns and that, consequently, the heterosporous water ferns belong to other main 'lines' of the Cormophyta, *viz.*, to those of the Pteridospermopsida. The often strikingly close resemblance between the fronds of Pteridosperms and true ferns, so troublesome in the palaeobotany of the Carboniferous, is a clear case of 'levels': the fronds of both groups developed independently in different lines from a branched telome system by the same morphogenetic processes of overtopping, planation and webbing (and provide a good example of parallelism, incidentally).

BECK's recognition of the Progymnospermopsida as the common group of progenitors of all spermatophytic plants has also been a step in the right direction. His suggested over-all phylogenetic lineages, leading to the Coniferopsida (Coniferophytina) and Pteridospermopsida, respectively, corroborates the generally held views of taxonomists and palaeobotanists that these groups show certain relationships but must have separated early in their phylogeny. As I have pointed out before (MEEUSE 1963b), BECK's suggestion can be extended by linking up the evolutionary lines and semophyleses in the Coniferopsida worked out by FLORIN with a common progymnospermous ancestral group, and by postulating a common morphological prototype for the seed ferns, which can in turn be used as a starting point of other semophyleses. It also clarifies the relative position of various other groups, *e.g.*, refuting any close relationship between *Ephedra* and the Coniferopsida, as recently as 1952 still defended by EAMES—they belong to different 'lines' already separated in the Lower Carboniferous. These instructive corollaries of a fairly simple basic assumption and the mutually corroborative evidence from various sources resulting in sudden advances, to my mind, represent positive and lasting achievements of the New Morphology, of what LAM calls 'the dynamic approach', and inspire confidence that we are on the right track and may eventually succeed in raising phytomorphology from its state of static lethargy.

9
Axial and Lateral Organs

In the Classical Morphology, which is 'Angiosperm-centred', the axial or cauline and lateral or foliar organs were supposed to represent static, existing ('given') and mutually exclusive categories. The morphological nature and the homologies of both fertile and sterile organs were all explained in terms of 'stems' and 'leaves' (the 'axial' and 'lateral' organs, respectively), the 'roots' forming a special category of axial organs distinguishable from the proper 'stems' by some characteristic features such as the lack of leaves, a different mode of branching, a different stelar anatomy, etc.

A tacit and not often explicitly expressed consequence of this conventional classification of all organs is that in the normally aerial portions of the plants only monaxial systems of cauline organs can exist, each axis being the axillary derivative of the lateral phyllome, so that each ramification of the axis can be established only by the development of axillary shoots from the axils of the appendicular leaves. It is true that in the vegetative regions of the Higher Cormophyta, especially of the dicotyledonous Angiosperms, the distinction between 'cauline' and appendicular organs is a sharp one, even if the leaf–axillary shoot relation is not evident. Nevertheless, even the traditionalists had to compromise by recognising other forms of branching, such as sympodial and di- or trichotomous ramifications, especially in the Lower Tracheophyta. BOWER was probably the first to express some doubt about the distinction of 'leaf' and 'stem'. He arrived at this conclusion through his fern studies, and indeed one only has to observe such forms as *Lygodium* to understand that the categories of 'stems' and 'leaves' are in some groups not at all clearly defined. H. POTONIÉ, HALLIER, TANSLEY and especially LIGNIER gradually extended BOWER's deductions, to a large extent after studies of fossil plants (see R. POTONIÉ 1959). Their ideas finally culminated in ZIMMERMANN's telome theory, which was largely inspired by the discovery of the *Rhynia* flora (KIDSTON and LANG) and simply postulates that the axial and lateral organs are derived from a common phylogenetic prototype, the telome, by a divergent semophyletic differentiation of groups of telomes. In a historical review, LAM (1948) has pointed out that, although such a comprehensive explanation has given us a clearer insight into many phylogenetic problems (see also Chapter 7), it is impossible to give a satisfactory and unambiguous static 'definition' of 'stems' (axial or cauline organs) and 'leaves' (phyllomes or appendicular organs). It is still mainly the relative position in respect of other organs that serves as a morphological criterion to refer an organ to either class, the *functional* differences being inadequate; and this leads to the somewhat paradoxical situation that a morphological entity, derived from a certain prototype, is regarded as 'axial' in one evolutionary line and as 'lateral' in another. As an example: the lateral segments of the first order of the sterile frond-like complex syntelomes of the Progymnosperms (*Archaeopteris, Aneurophyton,* etc.) consisting of a central 'rhachis' bearing lateral assimilatory organs ('sterile pinnules'), developed into a vegetative branchlet (a shoot, often a stunted one or 'brachyblast') in the Coniferopsida, the rhachis being represented by its semophyletic derivative, the 'branchlet', 'twig' or 'shoot axis', and the lateral pinnules by the appendicular assimilatory organs, still the *functional* phyllomes or 'leaves'; whereas the *whole* progymnospermous frond semophyletically evolved into the vegetative phyllome of the Higher Cycadopsida, which has incorporated the ramified axes and the assimilatory pinnules of the archetype and appears as a single lateral organ in respect of the axis (a main 'trunk' or at least a major 'branch') that originally supported the complex frond. To make things even more complicated, the anatomical structure of such a complex progymnospermous (and lyginopterid or medullosan) frond shows much resemblance to that of a 'stem' (*e.g.,* by exhibiting secondary growth) in the basal petiole-like portion ('stipe') of its rhachis, considerably less so in the ultimate ramifications. One can say that, roughly speaking, the frond anatomy becomes less 'cauline' and more 'foliar' as it is farther removed from its point of attachment. This explains why, in many cases, the anatomy is not always adequate in deciding whether an organ is 'axial' or 'lateral', especially in palaeobotanic studies of fragmentary fossil remains. In view of possible semantic complications, extreme care must be exerted in the interpretation of such structures in terms of the old-fashioned phytomorphology.

Another possible distinction between 'leaf' and 'stem' may rest upon a difference in 'function'. It is true that as a rule the vegetative phyllomes have an assimilatory and the stems a conducting and supporting function, but there are many exceptions, chiefly among the Flowering Plants (such as the assimilatory caulomes of stem succulents and various cladodically transformed axial elements). This again is not very helpful. Other non-phylogenetic criteria have been suggested, but to no avail, and it is not surprising that several phytomorphologists, more or less in desperation, concluded that the difference ultimately escapes us. I shall not discuss MAEKAWA's expositions on the subject, because his concepts of 'leaf' and 'stem' are most unusual and do not contribute to any tangible improvement. ARBER's (1950) 'partial shoot' theory of the leaf is another example of confusing connotations; a lengthy discussion of her views in phylogenetic botany would not serve a useful purpose.

Is the position really so bad that we must abandon the notions or concepts of 'leaf' and 'stem', of 'axial' and 'lateral' organs, altogether? One must first agree upon the question of a possible homology between sterile (purely vegetative) and fertile (sporangiate or reproductive) organs. The conventional phytomorphological tenet—that only three basic categories of organs, *viz.*, leaf, stem and root, exist—necessitates the inclusion of the fertile organs in one of the three; since the root category obviously does not qualify, it becomes a choice between an axial or a lateral interpretation. This moot point will be discussed in a different context in several other chapters (especially in Chapter 12), and it suffices for the time being to mention the fundamental postulate, accepted by the present author, that only the sporangia and their semophyletic derivatives are organs *sui generis* in respect of all phylogenetically sterile organs. One can trace the vegetative organs to their semophyletic origin by defining them as derivatives (homologues) of other morphological entities such as syntelomes or telomes, so that the question of what category the sporangiate organs belong to is decided solely by the semophylesis of the sterile elements ('sporangiophores', 'fertile telomes') on which the sporangia are borne. However, this does not mean that we must at all cost force the reproductive organs into the straitjacket of the conventional or 'classical' categories of stem and leaf.

A second consideration is the possible conceptual and semantic confusion of the term 'appendicular'. In the 'monopodial' (*i.e.*, monaxial) system of branching in the Flowering Plants an 'appendicular' organ is of necessity a 'leaf', but in an 'overtopped' (*i.e.*, anisotomously bifurcate) syntelome the overtopped lateral ramifications, although they are 'appendicular' cannot be called 'leaves' (phyllomes). Even the much more advanced complex fronds of the Progymnosperms and the seed ferns are polyaxial systems (there are no bracts!) of which at least some of the ultimate ramifications, although lateral in respect of the supporting axis, are not unequivocally foliar by any standard. The telome theory explains the origin of the two categories of vegetative organs out of a postulated, originally isotomous and repeatedly bifurcating branching system by assuming different developments and different relative positions for the partners of a bifurcation. The qualification 'phyllome' is applicable only to those syntelomes or portions of syntelomes that are manifestly laminose (and essentially assimilatory) organs. The overtopping and hence supporting elements remain non-foliar and must belong to the alternative category of the cauline or axial organs irrespective of their position in respect of other caulomes. In the progymnospermous taxa, the ultimate segments of the frond-like complex syntelomes were partly assimilatory lateral organs and, at least functionally, 'leaves', partly slender axes bearing a number of 'fertile pinnules' with stalked sporangia which, in spite of their basic homology and topological correspondence with the foliar pinnules of the same frond, are non-foliar (see page 24). The question as to whether in the Higher Cormophyta the sporangia or sporangial homologues are 'leaf-borne' or 'axis-borne' is thus phylogenetically solved in that the sporangia were initially borne terminally on telomic sporangiophores inserted on the rhachis of a fertile pinnule which has no specific attributes of a phyllome, and not marginally on a 'leaf'. The fact that the sporangiophores are 'appendicular' or 'lateral' in respect of the rhachis of the fertile pinnules does not make these terminal telomes parts of a 'foliar' organ.

Once a syntelome has become lateral in respect of a supporting axis and assumes the characteristics of a flat assimilatory organ, it is a functional phyllome. The functional trophophylls of the Coniferopsida are the homologues of progymnospermous lateral assimilatory organs and would qualify. However, this can be endorsed only if the homology of these organs with the lateral assimilatory syntelomes of other groups is not considered to be a major requisite of the concept 'leaf'. The functional trophophylls of the cycadopsid groups—and these include the leaves of the Angiosperms—are the homologues of whole vegetative progymnospermous 'fronds' and developed by the progressive increase and subsequent coalescence of the surface area of the assimilatory elements (the 'laminae') of the pinnules; the axial 'skeleton' of the pinnae and, ultimately, of the whole frond became incorporated as the 'veins' and the 'principal veins' ('midrib' or 'costa', etc.) of the simplified leaf (ASAMA 1960, 1962; MEEUSE 1963b). The midrib of the ultimate phase, the simple leaf, originally formed part of an elaborate branched system of axes, but after its incorporation in the blade it is, without previous knowledge of its semophyletic history, no longer recognisable as an axis bearing subsidiary caulomes because it forms an integral portion of a single 'leaf'. Conversely, the main rhachis of the sterile pinnae of a progymnospermous frond became promoted to the status of a trophophyll-bearing branchlet in the Coniferopsida, i.e., of a caulome (a 'stem'!) bearing lateral 'leaves'. Such relations emphasise the correlative and mutual definition of the functional 'stems' and 'leaves', in other words, of the relativity of these concepts.

The evolutionary changes that began with a simple overtopping process and ultimately resulted in various complexes of functionally caulomic and phyllomic organs also had a far-reaching effect on the development and growth of the branched system of axial and assimilatory structures that constitute the progymnospermous and pteridospermous fronds, the coniferopsid brachyblasts and the cycadopsid leafy shoots. In the most primitive psilophytic condition of isotomous bifurcation (dichotomy), the telomes developed from an apical growing centre, possibly from a single apical cell, the bifurcations initiating from an equal division of the apical meristematic cell. Each half grew out into a telome which was practically identical with its counterpart. Anisotomy (overtopping) can originate only from an anisotomous division of the apex and the increasing predominance of the overtopping (axial) element. The semophyletic evolution must have included a tendency towards a helical arrangement of the lateral elements by an increase of the number of orthostichies, and gradually the growing point must have assumed the character of a self-perpetuating assembly of meristematic cells forming lateral primordia at regular intervals—but the phylogenetic history of the shoot apex is a very insufficiently known and indeed much neglected subject.

An important innovation was the development of the axillary shoot (bud)–axillant phyllome (bract) relation. In the more primitive vegetative shoot apices, which produce only foliar (lateral) elements without axillary derivatives, the branching of the stem initiates as a division of the apex itself and the branching of the stem is of necessity dichotomous (rarely trichotomous) with isotomy (trichotomy) of the bifurcations, or, if one of the secondary branches predominates, of a 'sympodial' type of growth. Isotomous ramification is quite common among such primitive groups as Lycophyta (Lepidodendrales!), but it is also the predominant type of branching in roots and, significantly, still in the stems of caulescent Monocotyledons such as some Pandanales, Arecaceae, Bromeliaceae, Dracaenoideae, 'Agavaceae' and *Aloë*, a habit which is associated with the absence of axillary buds in the vegetative region of the majority of the Monocots and is undoubtedly a primitive cycadopsid feature. Sympodial modifications of this form of branching seem to have been present in some Bennettitalean taxa (Williamsonieae). In several main (phylogenetic) lineages, vegetative axillary buds developed, a situation which is concomitant with the advent of the monopodial type of branching as found in at least some Coniferopsida, *Gnetum,* and practically all stem-forming Dicotyledons (the 'secondary' type of monopodial branching among the Dicots by the cessation or suppression of the activity of the original shoot apex, the function of which is taken over by a lateral bud, is of course a fundamentally different form of ramification).

The origin of the vegetative axillary bud is another much neglected subject. The 'unit' consisting of a trophophyll and its axillary bud originates histogenetically as a single lateral primordium which subsequently forms an adaxial bulge that becomes the bud (shoot) primordium. The vascularisation of a bud and its subtending leaf, which is normally based on a common vascular trunk, also indicates that these two elements form parts of some organic entity, so that the phylogenetic interpretation must be based on the assumption that the leaf and its axillary bud were elements of a complex structure that contained foliar and cauline (or rather prefoliar and protocauline) components. The evolutionary trends within each particular group eventually determined which cases developed axillary buds, whether that complex structure may have been something of the nature of a progymnospermous vegetative frond or its unsimplified homologues (coniferous short shoots), or a cycadopsid leaf-bearing main branch. In either case, the axillant foliar organ must be the most proximal trophophyll of the vegetative branch that shifted to the axis supporting the branch, or a stipular differentiation of the petiole-like stipe of a complex frond (such stipular differentiations already occurred in *Archaeopteris;* see BECK 1962). The original bifurcate mode of branching suggests that during the semophylesis the anisotomy became very pronounced, so that the lesser ramification became strongly overtopped and distinctly lateral. A retardation of the development of the shoot in respect of the axillant basal phyllome is all that is required to produce the axillary bud. One could also visualise the retardation as an adaption to severe winter conditions, but that does not make these deductions any less conjectural. There are as yet no unequivocally corroborative palaeobotanic data, but it is quite clear that the vegetative axial bud, a condensed lateral shoot consisting of an axis bearing functional trophophylls, could only develop from a situation in which large progymnospermous compound 'fronds' differentiated into a system of coaxial twigs (in the Coniferopsida) or into simple leaves (in the Higher Cycadopsida), the biaxial stem-lateral branch system being the primary condition, the subtending phyllome only the secondary phase.

The advent of the axillary bud had far-reaching consequences involving the whole gross morphology and the ecology of the plant, such as the habit, the formation of a large crown of leaves, the rapid regeneration of the trophophylls after simultaneous shedding of the leaves at the beginning of a dry season or a cold winter followed by a period of dormancy, and the facilitation of the secondary growth of the stem, the leaves being formed on the new 'annual' growths.

The relation between a phyllomic 'bract' and a fertile (sporangiate) axis, so common among the Higher Cycadopsids, must, although it is comparable with the association of vegetative buds (shoots) and trophophylls (because the semophyletic origin is presumably also a segregation of originally coaxial elements by a shift of a proximal phyllome to the subtending axis of a lower order; see MEEUSE 1963b), be an independent development. Significantly, bracts are almost invariably present in the floral regions of monocotyledonous taxa which normally lack vegetative axillary buds.

For the purpose of definition, the functional trophophylls of all Spermatophyta (which are not all homologous entities, as we have seen) can be related to the syntelomic complex fronds of the ancestral Progymnosperms. For simple reference purposes, one may call all these assimilatory organs and their homologues, collectively and individually, 'leaves', 'phyllomes', 'trophophylls' or 'lateral (foliar, appendicular) organs', as long as the semophyletic (phylogenetic) relations are clearly understood and possible confusions are precluded. However, one must also be aware of the danger of the unrestricted homologisation of all forms of phyllomes, even of those of a single individual, such as normal assimilatory trophophylls, bud- and rhizome-scales, various bracts, cataphylls, prophylls, perianth lobes and cotyledons. Even if one accepts the fundamental homology of these organs, their morphological and anatomical characteristics may be so different that a comparison is meaningless. The frequently used argument that the vascular anatomy of the cotyledons represents the phylogenetically primitive ('original') vascular pattern of all vegetative leaves and all floral appendages (perianth lobes, carpels) is based on the assumption that the various classes of phyllomes of a single individual are strictly homologous, but this is of course not true. The cotyledons of the Dicotyledons must be the direct derivatives of the already differentiated *cotyledons* of some gymnospermous progenitor, which was presumably still at a protocycadopsid level of organisation, and the same relation exists between the *trophophylls* of the Angiosperms and those of their Mesozoic progenitors. The independent semophyleses of the various categories of phyllomes render a close correspondence between their anatomical features at the present-day level rather dubious. That cotyledons (and cataphylls) are indeed anything but 'identical' with trophophylls is demonstrable, among other things, by their different phyllotaxis and by the lack of axillary buds in the axils of the cotyledons of the Dicotyledons.

A collective term of reference ('trophophylls', 'phyllomes', 'leaves', etc.) could also be applied to the assimilatory organs of the sporophytes of other cormophytic groups in which the plant body is differentiated into rod-like (*i.e.*, functionally at least, 'axial') and laminose 'appendicular' elements, with the same restriction, *viz.*, only if in the relevant context the intention is clear and unambiguous. I feel that, to avoid undesirable connotations, one can and should employ such terms as a 'Lyco-leaf', a 'fern-frond', the (functional) leaf (or trophophyll) of the Coniferopsida, and, in a pinch, could thus even speak of the 'leaf' of the 'foliose' bryophytic gametophyte, but not of 'the leaf', *tout court*.

The 'sporophyll' concept postulates the existence of a category of foliar organs which are supposed to be the homologues of sporophylls but to differ from the normal vegetative phyllomes in that they are fertile. The tenacious adherence to this archaic axiom leads to such extreme and untenable consequences as the ideas expressed by EAMES (1961) concerning the *de novo* origin of the angiospermous ovules as trichomatic enations, rightly criticised by CAMP and HUBBARD (1963a). Neomorphological principles require that homologous organs have a common phylogenetic origin and, as we have seen in Chapter 3 (page 21), a derivation of trophophylls from 'webbed' planated aggregates of sterile telomes and of trophophylls from similar syntelomes consisting of sterile and fertile (sporangiate) telomes was proposed by ZIMMERMANN. The homology rests primarily on the basic morphological identity of all telomes, but it requires the proviso that the fertile telomes also underwent the morphogenetic processes of planation and webbing. This may have been the case in the early phases of the semophylesis of the sporangiate fronds of the true ferns, but the palaeobotanic evidence does not confirm the occurrence of 'webbing' of fertile telomes (sporangiophores) during the early evolution of other groups of vascular plants. Another difficulty is that the formation of a laminose organ from an association of sterile and fertile telomes could conceivably only produce a 'sporophyll' bearing marginal *sporangia*, whereas the Spermatophyta are characterised by the fact that they bear *ovules*. The homology of the nucellar part of an ovule with a megasporangium and the homology of all ovules can be postulated, so that the semophylesis of the ovule and its protective coverings (integuments, cupules) can be reconstructed (see Chapter 15), a procedure which necessitates the selection of prototypes in which the ancestral sporangia (or their sporangiophores, the 'fertile telomes') were borne on telomic axes that never primarily formed a laminose lateral organ with the attributes of a trophophyll. Secondary semophyletic associations of the sporangiate axes or their fertile homologues with phyllomes in structures more or less answering to the preconceived idea of a 'sporophyll' are of course irrelevant in this connection, because the new structures are not the homologues of a leaf, but of a combination of a vegetative (sterile!) foliar organ and a fertile cauline or precauline complex, the sporangial homologues remaining axis-borne. In the Spermatophyta, at least, *the classical 'sporophyll' simply does not exist.*

One aspect that has apparently not been studied systematically is the connection between the phylogeny of the lateral and axial organs and the evolution of the shoot apex from a dichotomously splitting apical cell to a complicated angiospermous growing point with its many layers of meristem cells differentiated into a 'tunica' and a 'corpus' and forming primordia in a regular sequence of plastochrons. This subject is intimately connected with the old problem of phyllo- and stachyotaxis, but a phylogenetic approach has so far hardly been attempted, as far as I am aware. For the time being only some general suggestions can be made. The basic, dichotomously ramified, telome systems of many Eocormophyta branched in a few constant directions, which implies that the primary orthostichies are 'older' than the classical 'genetic spiral'. Overtopping and planation of the bifurcating telome system did not fundamentally change the situation, the 'overtopped' ('lateral') syntelomes remaining inserted on the overtopping protocaulome in 2, 4 or occasionally 3 (or 6) vertical rows ('orthostichies'). The resulting stachyo- and phyllotactic relations could only be of the simplest kind, only angles of divergence of 90°, 180° or 120°, corresponding with 4, 2 or 3 orthostichies being present, apart from some verticillate types with whorls of 2 or 3 lateral syntelomes at each node (an overtopped trichotomously branched system would produce a kind of decussate phyllo- or stachyotaxis from which, in the planated form, a distichous arrangement would result). It is conceivable that some simple phyllotactic patterns such as occur in chlamydosperms (*Gnetum*) and various more primitive angiospermous groups (Piperales, Dilleniales, some of the Polycarpiceae and Monochlamydeae) represent such an ancient condition.

The much more complicated phyllotaxies of many Higher Spermatophyta, such as Cycadales, Bennettitales and several orders of the Monocotyledons, must have originated from these few simple, basic types. One of the possible ways is the 'torsion' of each orthostichy into a helix by the constant 'shift' of each of its leaves by a few degrees in respect of the leaf below it, a condition which is conspicuously demonstrated by the 'screw palms' (*Pandanus*) and several other forms (*e.g.*, some Zingiberales). In other words, the original (primary) orthostichies became helical parastichies and some new radially symmetric patterns developed. The secondary spatial superposition of leaves belonging to the different parastichies thus formed would result in new orthostichies whose angles of divergence are determined by the shift of the leaves that formed the primary orthostichies (*i.e.*, of the 'steepness' of the helices formed). In this way various more complicated phyllotaxies would emerge. Although much more research is needed, there is a fundamental agreement between these phylogenetic deductions and the ideas developed by PLANTEFOL (1946–1947), who also explained phyllotaxis as a radially symmetric system of parallel helices after his developmental studies of shoot apices. However, several angiospermous taxa exhibit inconstant phyllotaxies, the phyllotaxy sometimes varying in a single specimen or changing with the ageing of a plant, so that another and alternative form of 'shifting' from a phyllotactic pattern to another one can actually be observed, *viz.*, a re-arrangement of the elements of a *single* helix (the 'genetic spiral'). From the primitive patterns corresponding with the phyllotaxies in the ratios 1/2 or 1/3, a phylogenetic shift of the leaves of a single phyllotactic helix may also have resulted in the more complicated orthostichies with angles of divergence of 144° (2/5), 135° (3/8), ±138° (8/21), etc. The reasons or causes underlying the shifts are probably a tendency of the plants to arrange their assimilatory organs in such a fashion that the maximum surface area is exposed, a tendency which, at least in shoots with rather crowded phyllomes, would lead to an increase of the radial planes of symmetry. A recent paper by PHILIPSON and BALFOUR (1963) is interesting in this connection, as it suggests a line of inquiry based on the relation between stelar anatomy and phyllotaxis.

Finally, a few lines must be devoted to the terminal or pseudo-terminal position of several organs. Some authors flatly deny the occurrence of terminal ovules, carpels and leaves (in Angiosperms) and restrict the truly terminal elements to the growing points, but the terminal position of the sporangia of Psilophytes (and even of Progymnosperms!) implies that sporangial homologues (ovules, thecae) are essentially still terminal on their stalks (synangiophores, funicles, filaments). The relative position defining the cauline and foliar character of vegetative organs, the terminal portions of a complex syntelome when forming a trophophyll of some sort may conceivably even produce a terminal 'leaf'. Quite apart from these considerations, the factual (pseudo-) terminal position of many elements such as (carpellate) gynoecia and ovules renders many discussions rather inane. If one feels inclined to refer to an organ in a topologically terminal position as a 'terminal' organ, for instance, when the term is used descriptively or the element 'behaves' (*e.g.*, physiologically) as a terminal organ, the application of the term 'pseudo-terminal' serves no useful purpose, especially if the morphological interpretation is not definitely settled. The frequent and sometimes hairsplitting disputes concerning the terminal or lateral ('appendicular'!) position of ovules, which are the starting point (or the outcome, as the case may be, because quite often it is simply a matter of circular reasoning!) of different interpretations of the gynoecia (phyllospory, stachyospory, etc.), has no demonstrative force (MEEUSE 1964b). Phylogenetically speaking, ovules and androsynangia (staminal thecae) may be either laterally or terminally borne on a supporting organ (synangiophore, cupule, gonoclad, etc., but not on a leaf!), depending on the case under discussion, *i.e.*, on the divergent semophyletic evolution of the sporangiate organs in the Higher Cycadopsida. A similar case is the much-debated position of the single cotyledon of the Monocotyledons. SWAMY (1963) has clearly shown that the concepts 'lateral' and 'terminal' are inadequate to describe (and to distinguish) the condition of monocotyly, because the single cotyledon, as a derivative of a sagitally divided terminal cell, is not lateral in respect of the hypocotyl, but not truly 'terminal' either, as it develops from one of the two daughter cells of the apical cells and has a counterpart in the epicotyl (it is, in fact a kind of dichotomy of the apical cell of the proembryo). In phylogenetic inquiries into problems of homology and of morphological interpretation, the question of a terminal *versus* a non-terminal insertion of elements is no longer a major issue and has become almost completely superfluous.

10
Phylogeny of the Reproductive Region: I. General Considerations

The reproductive structures of Lycophyta, Sphenophyta (Articulatae) and the true ferns will be left out of consideration here because their semophyleses are, upon the whole, fairly clear if one disregards the indiscriminate use of the term 'sporophyll' for the sterile lateral organs (bracts, stegophylls) subtending or bearing sporangia. The publications by LAM (1948 *et seq.*) and by ZIMMERMANN (1959) may be consulted for details. One should bear in mind that these groups bear sporangia or sori, so that any attempt to relate the ovuliferous fertile organs of the Higher Cormophyta to the sporangiate organs of any of these pteridophytic groups must include a plausible explanation of the origin of the characteristic accessory organs of the ovule, such as the integuments.

The reproductive organs of the Angiosperms, on the other hand, are the most hotly debated structures in phytomorphology. The Gymnosperms, as their next of kin, became involved by their unmistakable phylogenetic relationships with the Flowering Plants, so much so that their reproductive organs were often considered to be homologues or even semophyletic prototypes of the genitalia of the Angiosperms. Examples are the so-called 'megasporophyll' of *Cycas* and the 'microsporophyll' of *Cycadeoidea*, which have repeatedly been compared with the carpels and the stamens of the Ranalian Angiosperms, respectively. The quest for the nearest gymnospermous relations of the Angiosperms, which might even qualify as their putative ancestral forms, was until recently hampered, among other things, by the lack of understanding of the relationships between the various conventional groups of the Gymnosperms themselves. In the last decades, the idea had been gaining ground that the Gymnosperms constitute a formal but extremely heterogeneous assembly of not very closely related (or even unrelated) major taxa. When BECK (1960) proposed the name Progymnospermopsida for a

number of mainly Upper Devonian fossils which, as the name implies.
are supposed to form the basic taxon of all gymnospermous groups (and,
I may add, of the Angiosperms), I believe he made a fortunate choice.
The postulation of such a group of still more or less 'pteridophytic'
progenitors of the Spermatophyta elucidates the heterogeneity of its
gymnospermous descendants (which is manifestly the result of ancient
divergencies of early gymnospermous phylogenetic lineages), whilst
showing their fundamental homology through propinquity of origin. It
thus becomes possible (see MEEUSE 1963c) to relate all gymnospermous
taxa to a common progymnospermous morphological prototype by retracing their differences in phenetic characters to divergent phylogenetic
trends and their correspondences in features to a common descent from
the same ancestral condition (a few probable parallelisms excepted).
The most likely semophyleses of both the sterile and the fertile organs
of the Spermatophyta can be construed, as will be shown in the following chapters.

The deductive method of the New Morphology, by linking the floral
morphology of the Flowering Plants with the structure of the reproductive
regions of the Gymnosperms, provides a new interpretation of the floral
region, i.e., of the majority of the traditional 'flowers', based on a polyaxial
morphological prototype (to be referred to as an 'anthocorm'). This
novel approach has certain consequences which sometimes complicate
and sometimes simplify the interpretative morphology of these fertile
structures, but its greatest advantage is that it aligns the floral morphology of the Angiosperms with that of the other cycadopsid groups. The
suggestion that the Angiosperms descended from cycadopsid Gymnosperms is by no means new or startling, but till now traditional doctrines
prevented phytomorphologists from exploiting this postulated phylogenetic relationship to the full. The villain of the piece is the 'sporophyll'
concept, from which followed the retrograde deduction that the ancestors
of the Angiosperms had also been phyllosporous. The proper method of
inquiry is to derive the reproductive regions ('flowers') of the Angiosperms semophyletically from—preferably discrete—gymnospermous
prototypes ('which had not yet become flowers', to use an Angiosperm-
centred phrase!). However, R. VAN WETTSTEIN's attempt to explain the
origin of the angiospermous flower, although it was based on the
morphology of the genitalia of a discrete gymnospermous prototype, also
failed because Angiosperm-centred connotations still led to muddled
thinking. His hypothesis did, in my opinion, come fairly close to the
solution of the problem, but the application of conventional terms caused
a clash of semantics and definitions (see Chapter 3). The primary
gynoecial element in WETTSTEIN's pseudanthium is the chlamydote ovule
of a gnetalian form, traditionally but quite erroneously called a 'female

flower'. A flower, in the conventional definition, is a modified shoot (brachyblast) bearing, among other things, *carpels* with essentially marginal ovules. In other words, conceptually there can be no (female) flower without at least one carpel, so that, to qualify as a 'flower' in the classical sense, the female reproductive organ of a Chlamydosperm must also have a carpel. The solitary terminal gnetalian ovule, surrounded by a chlamys, does not easily conform to the standard definition of a carpel as a laminose organ with several marginal ovules. HAGERUP (1934 *et seq.*) and others have identified an integument or the chlamys with (the sterile part of) the carpel, but this conclusion, manifestly inspired by the classical tradition that ovules must be leaf-borne, was decidedly forced. The alternative current interpretation of the chlamys, *viz.,* its identification with a 'perianth', is also absurd, because a perianth surrounds the genitalia *of a flower,* so that if a structure envelops no more than a single ovule it cannot be a perianth. The chlamys is apparently something different, but, quite apart from its nature and origin, it does not play a conspicuous role in the pseudanthium hypothesis anyway, because one of the essential requisites of any floral theory is that it must provide a plausible explanation of the origin of the Ranalian follicle, in which several ovules are associated with a single 'sporophyll' (carpel). Several protagonists of WETTSTEIN's hypothesis (KARSTEN 1918, and JANCHEN 1950) have attempted the derivation of such pluri-ovulate organs from the chlamydospermous gynoecium in which only a single ovule is contained, but their suggestions are entirely unconvincing. JANCHEN declares 'that it is easy to visualise' the formation of a Ranalian carpel out of aggregates of such solitary ovules (and, presumably, bracts providing the valvular sterile portions of the gynoecium), but at the same time maintains that the genitalia are leaf homologues. He is apparently caught in a strange labyrinth of (classical) definitions and semantics. The supposed semophyletic origin of the 'sporophylls' (stamens and carpels) from the 'cauline' microsporangiate organs and the solitary ovules of the chlamydospermous archetype clashes with the conventional circumscription of the genitalia, based on the postulate that the sporangia (or ovules) are initially leaf-borne. In the primitive pseudanthium, the ovules (the so-called female gnetalian flowers) are supposed to be inserted on the main axis of a complex structure that ultimately evolved into a flower, in other words, on the floral axis itself. The ovules, as presumed leaf-borne elements, must be attached to the carpel, so that one must assume a most unlikely 'shift' of the ovules from the main axis onto the carpel, a conclusion at variance with the classical doctrine of the primarily leaf-borne ovules.

NEUMAYER (1924) contributed some interesting emendations of the pseudanthium hypothesis but although his much neglected paper con-

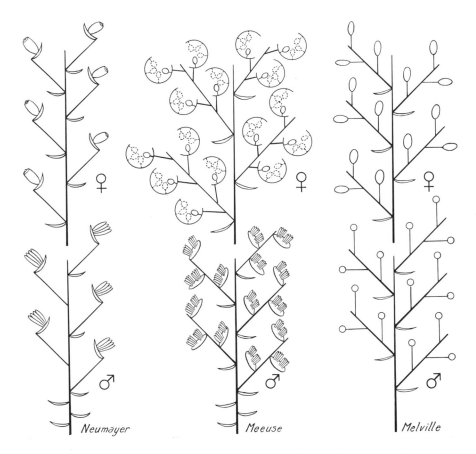

FIGURE 4. *Left*—Diagrammatic representation of an anthocorm in the sense of NEUMAYER, female version and male version (if the central axis is continuous, the ambisexual version). The central axis is supposed to bear bracteated gonoclads (andro- and gynoclads), each gonoclad bearing a single 'sporophyll' with a single ovule or a single anther. *Centre*—AUTHOR's version of a primitive anthocorm, the central axis bearing bracteated gonoclads, the gynoclads bearing a number of ovuliferous cupules, the androclads a number of androsynangia on a flat supporting organ (male analogue of cupule). *Right*—A compound structure bearing 'gonophylls' in the sense of MELVILLE, each 'gonophyll' being supposed to consist of a bract and a sporogenous axis. The sporogenous axes are supposed to be bifurcate in MELVILLE's theory, but they are drawn as single axes for easy comparison with the other figures. The main objection to his theory is that the advent of integuments, cupules (arils) and androsynangia is unexplained.

tains more sound ideas than he is usually credited with, he confused the issue by introducing several highly conjectural ancillary assumptions whilst retaining some of the least acceptable postulates of WETTSTEIN's original theory. I find that NEUMAYER's basic idea comes very near a general phylogenetic interpretation of the reproductive region of all Higher Cormophyta and that, after the necessary readjustments, the fundamentals of his theory and also his nomenclature can be accepted. As the basic type of the floral region NEUMAYER did not suggest an aggregate of elements which is manifestly intended to represent the prototype of an angiospermous flower (the pseudanthium), but a much more general structure, the 'Anthokormus' (anthocorm). Although he did not express his conception of an anthocorm in the form of a diagram, it is quite clear that it corresponds with the structure represented here in Fig. 4 (*left*). It consists of a central main axis bearing a number of bracts, in a helical or verticillate arrangement, of which all except the most proximal 'sterile' ones subtend a fertile axis or *gonoclad*. Each gonoclad was supposed to bear a vestigial 'sporophyll' with one ovule or a male synangium. I shall have occasion to point out some of the fallacies in NEUMAYER's theory in another chapter and only represent the emended version of a bisexual anthocorm here (see Fig. 4, *centre*). The fundamental difference is the postulation of completely axial gonoclads without foliar appendages (sporophylls), each with initially numerous male synangia ('stamens') or chlamydote (arillate) ovules. However, because it is only an improved version of a general archetype of the fertile regions of all Higher Spermatophyta (as NEUMAYER intended it to be) and considering that the term 'gonoclad' (or its male and female form *andro*- and *gynoclad*) is still appropriate, his nomenclature can be retained.

This brief mention of the structure of the anthocorm, in anticipation of the much more detailed discussion in some of the following chapters, only serves to demonstrate that indeed the phylogenetic interpretation of the floral region results in a discrete phylogenetic and typological archetype that links Higher Gymnosperms with the Angiosperms and can be described in a neutral terminology which is not Angiosperm-centred. A rather concise survey of the reproductive regions of the Gymnosperms in the following chapter shows that the morphology of the anthocorm can be related to that of progymnospermous 'mixed fronds'. Before the interpretation of the angiospermous fertile region is given in terms of anthocorms (instead of 'flowers'), other floral theories will be critically discussed, in the hope that the gains and advances emanating from the alternative approach of floral morphology will stand out even more clearly.

11

Phylogeny of the Reproductive
Region: II. The Gymnosperms

All semophyletic relations proposed in this chapter can be deduced from the postulated origin of the Spermatophyta, *i.e.*, of all conventional gymnospermous groups and, indirectly, of the Angiosperms, in the palaeophytic Progymnospermopsida *sensu* BECK (1960, 1962), provided one recognises an early divergence of the derivatives of the Progymnosperms into two main taxon phylogenies, one of which soon again broke up into two. The separation of these principal phylogenetic lines, which most probably was already completed in or just before the Lower Carboniferous, and the prevalence of divergent evolutionary trends in each of these lineages are reflected in the striking morphological diversity of the assembly of gymnospermous groups. A comparison of all these major taxa, including both surviving and completely extinct representatives, can only be based on their fundamental homology by propinquity of descent from a common Devonian progymnospermous archetype. Similarly, the reconstruction of semophyletic sequences in each evolutionary line must also start from such an ancient prototype. A comparison of these semophyleses is nevertheless very instructive as an example of interpretative morphology, because it reveals certain primary phylogenetic relationships, such as the basic homology of all ovules and their integuments, whilst demonstrating the completely different evolutionary pathways taken by the individual groups. The independent semophyletic history of the most characteristic morphological features of each group explains why, at the present-day level, the gymnospermous groups appear as a heterogeneous assembly of seemingly unrelated forms rather than as a homogeneous alliance of intimately associated taxa, whilst exhibiting several incidental indications of a closer phylogenetic relationship.

The two principal ancient phylogenetic lineages rooting in the Progymnosperms ultimately developed into the Coniferophytina *sensu* ZIMMERMANN (Coniferopsida of many authors) and the Cycadophytina (Cycadopsida), respectively. The latter soon again must have evolved along two different lines, one group emerging as the true seed ferns or Cycadofilices of the Euramerican coal measures and the other one as the Pteridosperms of glossopteridalean alliance, chiefly confined to the old Gondwana floral region. The adaptive evolution of two of these three main lines, *viz.*, of the Coniferophytina and the Euramerican Cycadofilices, soon 'petered out', the over-all morphology of all survivors essentially retaining the level already attained in the Carboniferous. Only in the third phylogenetic line did important semophyletic changes take place which resulted in a sequence of evolutionary levels from the pteridospermoid glossopteridalean forms to the protocycadopsid, the cycadopsid and, ultimately, the angiospermous type of plant.

The Upper Devonian progenitors of the Gymnosperms must have been a fairly large and already varied group, but as a common basis of comparison a convenient archetype, more or less representative of the Progymnospermopsida as a whole, must be singled out. I have (MEEUSE 1963a) selected *Archaeopteris* as the first approximation of a prototype providing an average common pattern, especially because this taxon is the best and most completely known example which is at the same time manifestly not a derived but rather unspecialised form. As regards the terminology to be employed, there are some possible semantic complications and the terms adopted here are primarily intended for descriptive purposes. The complex syntelomes of the adopted prototype will be referred to as 'fronds', which are branched in a single plane (as in *Archaeopteris*—in other Progymnosperms such as the Aneurophytales they may have formed a three-dimensional structure with bi- or trifurcate ultimate segments). These fronds are provided with a main rhachis which is represented in the diagrams as unbranched but was in fact at least once dichotomously bifurcate in many (or all?) Progymnosperms and also in the majority of the derived pteridospermous fronds. The main rhachis (and its two principal branches) bore large segments or 'pinnae' consisting of a main axis (rhachis of the second order) and numerous lateral 'pinnules' or ultimate segments. According to BECK (1962), to whom we owe a neat reconstruction of *Archaeopteris*, in some forms the pinnae were interspersed by or alternated with small single sterile pinnules, but this was probably not a general feature among the Progymnosperms. It is possible that these solitary pinnules and the adjacent co-axial pinnae became associated at some later stage in the phylogeny of the Gymnosperms to form a kind of axillary shoot-bract relation. It is highly probable that the most proximal pinnules borne on the 'petiolar' basal portion of the main rhachis became aphlebia or other 'stipular' organs, or 'shifted' down to the main stem supporting the rhachis and later formed a 'bract' of the frond and its semophyletic derivatives. There were two types of fronds, completely sterile ones in which all the ultimate segments were vegetative (assimilatory), and 'mixed' ones which bore sterile and fertile pinnules on the same segment. A comparison of the morphology of the sterile and the 'mixed' pinnae indicates that there is some basic homology between the two, but as I have already explained (see page 24 and Fig. 1), this homology is based on the common origin from *telomes*, so that this relation cannot be extended to permit the identification of the homologous elements of the fronds with leaves (phyllomes), an identification which would make the sporangia leaf-borne. (Properly speaking, the conventional Angiosperm-centred terms and semantics are inadequate to cope with such situations at the syntelomic protocauline and protophyllomic evolutionary level, so that they cramp the style of the following deductions, which are of necessity somewhat forced into a traditional semantic pattern. However, I think I can still prove my point.) The sterile pinnules are syntelomes that evolved into protophyllomes *after* their semophyletic differentiation from the homologous protocauline syntelomes (the future fertile pinnules) which never became flat (laminose) assimilatory organs and hence can only be classified, 'by default', as axial structures. Not the future semophyletic development of an organ but its phylogenetic origin determines its morphological status. In this case, again, one must not confuse lines and levels!

As an ancillary argument, the vascular anatomy can be taken into account. The principal ramifications of the complex fronds of such forms as *Aneurophyton* apparently contained secondary stelar tissue and, considering that secondary growth is a feature consistently associated with 'stems' (caulomes), the rhachides of the ultimate pinnules, even though lacking secondary growth, are also rather of a protocauline (axial) than of a foliar nature. The elements of the fertile pinnules consist of stalked sporangia and their stalks or sporangiophores are clearly 'fertile telomes'. Accordingly, neither the sporangia nor their telomic stalks are leaf-borne but rather axis-borne, the sporangiophores being inserted on protocauline organs, so that the semophyletic derivatives of stalked sporangia such as ovules, funicles, and stamens are not primarily a part or an appendage of any organ that could qualify as a 'leaf' (CAMP and HUBBARD 1963b, MEEUSE 1963b). Secondary associations of sporangiate axial organs with foliar elements are not at all uncommon, as we shall see, but this is irrelevant in the present context.

The arrangement of sterile and fertile pinnules presumably varied among the Progymnosperms and did not always conform to the singular pattern found in *Archaeopteris* in which each 'mixed' pinna bears a number of distichously alternating sterile pinnules in its distal and proximal regions with a zone of similarly arranged fertile pinnules in between. It is not even certain that all Progymnosperms had both sterile and fertile ('mixed') fronds. Some of them (and of their pteridospermous derivatives) may have borne only 'mixed' fronds, but this is not particularly important.

The generalised prototype of a Progymnosperm (see Fig. 5), is supposed to be heterosporous, *i.e.*, it is assumed that the change-over from the initial (at least morphological) homospory to heterospory had already been completed. There is some evidence (see, *e.g.*, BECK 1962) that the mixed fronds of *Archaeopteris* were either microsporangiate or megasporangiate, in other words, that this taxon was diclinous. The rather consistent monoecy or dioecy in all Coniferophytina, in the Cycadofilices *s.s.*, and in the majority of the Protocycadopsida manifestly reflects their origin from progenitors with unisexual fronds (Coniferophytina) or at least with unisexual fertile pinnules (Euramerican seed ferns), but the established or suggested occurrence of bisexual structures in the Noeggerathiales, Marsileales and Glossopteridales indicates that in some Progymnosperms the fertile pinnules may have been ambisporangiate. In the following discussion of the evolutionary history of the fertile region, a unisexual (male or female) mixed frond will be used as the prototype, not only for convenience but also because in the majority of the groups (Higher Cycadopsida excepted) this was, and still is, the actual form of sex distribution.

It will become clear from the following chapters that such a complex progymnospermous mixed frond, as the semophyletic archetype of the fertile regions of all derived Spermatophytes, is the structure from which the general type of a floral region, the anthocorm with its gonoclads and various bracts, is derived, the 'pinnae' providing the fertile elements (andro- or gynoclads) and the sterile pinnae of the same frond the sterile 'foliar appendages'. It does not seem appropriate to refer to the archetype as an anthocorm with gonoclads, because if one defines a gonoclad as 'an axis bearing coaxial (stalked) ovulate cupules or male synangia' it represents a higher evolutionary level than the primitive mixed frond, although there was of course a gradual semophyletic transition. In addition, the gonoclads are normally subtended by bracts, and this gonoclad-bract relation must have become established before one can speak of an anthocorm.

CONIFEROPHYTINA

It must be postulated that the evolution of this group occurred in three evolutionary lines of descent which had already become independent in the Carboniferous and produced the three clearly distinct, recent taxa of the Ginkgoales, Taxales and Pinales (*s.l.*) The Cordaitales are considered to be morphologically more or less closely related to a common ancestral prototaxon of both the cone-bearing Conifers and the Taxales,

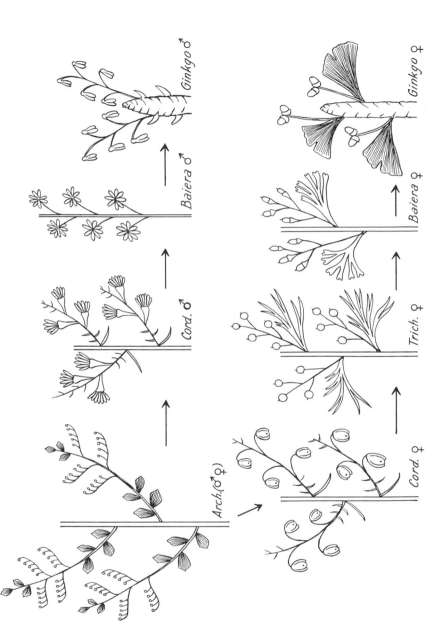

FIGURE 5. Tentative semophyleses of the reproductive regions of the Ginkgoales (diagrammatic): *Arch. = Archaeopteris; Cord. = Cordaitanthus; Trich. = Trichopitys.*

the structural pattern of their reproductive regions at any rate being cognate to a common archetype intermediate between the progymnospermous progenitors and their much more specialised younger descendants (MEEUSE 1963b). Morphologically, the most archaic surviving group is undoubtedly that of the Ginkgoales, which is corroborated by the fact that the maiden-hair tree is the only living representative of the whole class with zoidiogamy. Even the contemporary *Ginkgo biloba*, in spite of the oligomerisation of the number of ovules and of the number of microsporangia per synangiophore, can be directly related to a very ancient progymnospermous type. The most important semophyletic changes that took place are (1) the association of the sporangia of a fertile pinnule to form either an ovule or a stalked androsynangium, (2) the reduction of the sterile pinnules of the 'mixed' pinnae, although they are not completely wanting in the male (in *Ginkgo biloba* the distal ones are vestigial but often present in what could be called the protoandroclad, while the proximal ones are not rare as atavistic abnormalities; see Fig. 1), (3) the contraction of the compound mixed fronds and of at least some of the sterile fronds into a stout short shoot ('brachyblast') with a helical arrangement of the fertile axes and the leaves (or bracts), and (4) the development of a bract subtending each fertile homologue of a mixed pinna, which is presumably a secondary shift, to the axillant position, of some foliar element, *i.e.*, of a fertile pinnule of the main rhachis (the axis of the brachyblast) or of a pinnule of a rhachis of the second order (of a sporangiate axis) itself. The morphology of the more ancient precursors such as *Trichopitys* and *Baiera*, in conjunction with the frequent (atavistic) abnormal developments of the fertile axes of *Ginkgo* (NOZERAN 1955), allows the reconstruction of an evolutionary line (see Fig. 5) leading from the progymnospermous archetype to the recent *Ginkgo*. Apart from those advanced characters which are typically 'gymnospermous' (such as the bitegmic ovules), the features of the progymnospermous mixed frond are essentially retained, the functional leaves and bracts representing sterile pinnules, the ovules and stalked microsporangiate organs, fertile pinnules of the ancestral frond. The over-all morphology of the Ginkgoales may be taken as an indication of the early advent of the second integument.

The second line must have evolved into a pre-cordaitalean form, because I believe that both Taxales and Pinales, which are more closely related to each other than they are to the Ginkgoales, descended from an early type resembling the Carboniferous Cordaitales in many respects. One of the principal arguments against a *direct* derivation of the Pinales from the recorded cordaitalean forms, emphasised by such authors as FLORIN, is the coetaneous occurrence of Cordaites and early Conifers (Voltziales, etc.), but I do not think this is a serious objection. It is possible to relate the reproductive organs of both these surviving groups to a prototype which is essentially a more loosely constructed *Cordaitanthus* (MEEUSE 1963b). The recorded reproductive organs of the *Cordaitanthus* type formed rather dense structures (cones) and all that is necessary is the postulation of an earlier type, essentially a Cordaite, but with more primitive reproductive regions, here referred to as 'protocones' for easy reference. From this basic type, which is intermediate between the progymnospermous progenitors and the younger taxa, a tentative semophylesis of the genitalia of various groups can readily be deduced (see Fig. 6). The age-old problem of the female coniferous cone was virtually solved by FLORIN in a truly monumental and already classical series of studies (*e.g.*, 1951, 1954; see also LAM 1954). The diagrams of Fig. 6 demonstrate that FLORIN's semophyletic derivations can be extrapolated back into the past to link up with progymnospermous archetypes, which has the enormous advantage of allowing an interpretation of the female reproductive organs in terms of the morphology of primitive Devonian prototaxa instead of in the conventional (and confusing) Angiosperm-centred phraseology still marring FLORIN's papers (and NOZERAN 1955) and rendering them unduly cumbersome, apart from elucidating the origin and morphological nature of the corresponding male reproductive regions. A third and, to my mind, not inconsiderable advance is the manifest homology of the cordaitalean fertile 'brachyblast' (which later became transformed into the ovuliferous scale) with a segment of a progymnospermous 'mixed frond', so that all elements of such a cordaitalean brachyblast can be related to the various portions of the ancestral complex syntelome. The sterile scale-like foliar appendages of the brachyblast, variously interpreted as 'prophylls', 'bracts', 'scales', lobes of a 'sporophyll', etc., are derivatives of progymnospermous 'sterile pinnules', the stalked ovules and male synangia representing their coaxial fertile counterparts. The sporangia and their derivatives (ovules) are, accordingly, axis-borne rather than leaf-borne. A novelty was the advent of the bract (later becoming the coniferous bract-scale) subtending the female brachyblast. This bract can be visualised as having been a sterile pinnule secondarily 'shifted' to the axillant position. The semophylesis of the coniferous cone is concisely indicated by means of the diagrams of Fig. 6, a further discussion being unnecessary after FLORIN's scholarly disquisitions and various affirmative contributions by LAM, NOZERAN and the present author. The phylogenetic history of the male reproductive organs, also represented in Fig. 6, needs hardly any comment except for an interpretation of the morphological nature of the seemingly foliose microsporangiate organs, the traditional 'microsporophylls' of the Pinales. As semophyletic derivatives,

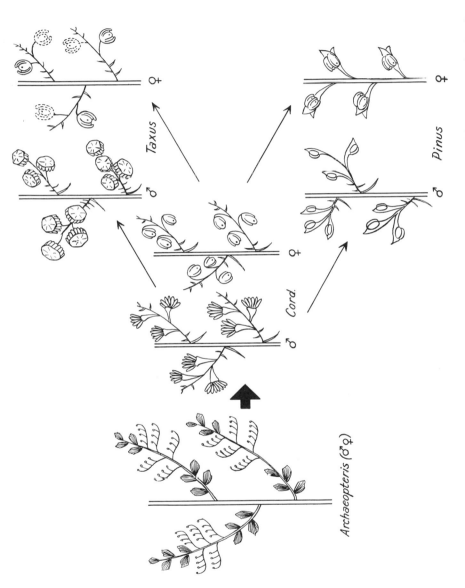

FIGURE 6. Tentative semophyleses of the reproductive regions of the Pinales and Taxales (diagrammatic) : *Cord.* = *Cordait-anthus.*

i.e., direct homologues, of male fertile pinnules, they are simply cladodically transformed stalked androsynangia and clearly axial organs.

The semophyletic origin of the female brachyblast also provides an
important clue to the interpretation of the organ which distinguishes
the gymnospermous ovule from older megasporangiate organs, *viz.,* the
outer integument. As I have explained elsewhere (1963b) the aggregation of the sporangia into dense clusters is a trend already conspicuous in
several progymnospermous groups such as the Aneurophytales. It is
likely that the megasporangia had already acquired an integument and
were in fact unitegmic ovules. The semophyletic identity of a solitary
stalked gymnospermous ovule with a progymnospermous fertile pinnule
bearing numerous megasporangia (or unitegmic ovules) tending to associate, like their male counterparts, into synangial complexes, leaves no option
but to accept with certain emendations BENSON's synangial hypothesis
of 1904 which explains the traditional integument of the lyginopterid
seed ferns (and by inference, of the Gymnosperms) as a concentrically
arranged group of co-axial sterile sporangia in juxtaposition surrounding
the only remaining fertile megasporangium of the original gynosynangium.
BENSON's explanation must be somewhat modified in that most probably
not megasporangia but unitegmic ovules coalesced into a synangial complex and that the sterilised ovules of the synangium formed the lobes or
segments that coalesced to form the outer integument (see also Fig. 10,
Chapter 14).

The third main coniferopsid genealogy, that of the Taxales *s.s.,* must
also be an ancient one (FLORIN 1954). Among the recent Coniferophytina, the New Caledonian *Austrotaxus* may well be the nearest approximation of the early Cordaitales; for although the principal trend in
the Taxales was the extreme oligomerisation of the number of ovules of
the megasporangiate mixed pinnae, the lack of any tendency to develop
female cones was responsible for the retention of an archaic (though reduced) protocone. A topological comparison with a postulated proto-
cordaitalean archetype reveals that the very short ovule-bearing axes of
the Taxales represent ovuliferous homologues of a progymnospermous
mixed frond, the fertile pinnules being reduced to one or two (rarely
more) subsessile ovules, and the sterile pinnules to a few scaly bract-like
appendages. The putative homologies (as well as those of the male
genitalia) are represented in Fig. 6. Characteristic is the strongly developed outer integument, usually called the 'aril'. The male synangia
are helically arranged around the supporting axis, which axis represents
the rhachis of a progymnospermous mixed pinna, each microsporangiate
organ being the homologue of a condensed fertile pinnule. In *Taxus* and
Austrotaxus these male synangia are terminal on their synangiophore and
peltate; in *Torreya* they are (apparently as the result of a secondary

reduction) unilaterally attached. These conditions can be visualised to have originated from a (pro-)cordaitalean stalked fascicle of micro-sporangia. The microsporangia were originally suberect to somewhat (and usually unilaterally) recurved. In the Pinales the fascicled micro-sporangia became completely deflexed, but they retained their individual-ity only in the rather primitive Araucariaceae, in practically all other groups the sporangia becoming adnate to their supporting synangiophore. In the ancient Ginkgoales (*Baiera*) and in the Taxales the microsporangia, unlike those of the Pinales, did not become unilaterally deflexed but assumed a radially patent to pendulous position. The lateral coalescence of the sporangia and the peltate distal enlargement of the synangiophore produced the characteristic microsporangiate organ of *Taxus*, which is clearly homologous with the microsporangiate organs (stalked male synangia) of the Ginkgoales and Pinales.

In contradistinction to all other Gymnosperms, the Coniferophytina possess discrete trophophylls (functional 'leaves') derived from *single* fertile pinnules of a progymnospermous frond. The pinnae of sterile fronds usually became vegetative short shoots or brachyblasts.

PTERIDOSPERMS

The recognition of the main phylogenetic trends in the fairly homo-geneous group of the Cycadofilices is comparatively easy, because a point-for-point comparison of a progymnospermous prototype (which was, presumably, rather of the *Aneurophyton* than of the *Archaeopteris* type) with the complicated fronds of these Euramerican seed ferns reveals that the latter essentially retained the gross morphology of the ancestral structures and that the advances consisted in the progressive aggregation of the microsporangia into synangia, the acquisition of the outer (synangial) integument of the ovule and the advent of the cupule. The fertile fronds of these Pteridosperms are truly 'mixed' in that sporangiate and vegetative pinnules are borne on a common supporting axis. In the more primitive lyginopterid groups the vegetative pinnules retained the narrow base of the archetype (cf. *Sphenopteris*), but in several lines there was a tendency towards a simplification of the fronds by the progressive coalescence of the ultimate pinnules into larger en-tities. This trend can be observed in various pteridospermous groups (ASAMA 1960, 1962). When the pinnules of a 'mixed' frond increased their laminal area, the co-axial fertile pinnules (transformed into cupu-lated ovules and peltate male synangia) became involved. They were sometimes 'bypassed' by the encroaching assimilatory elements, in which case they remained distinctly rhachis-borne, but occasionally the sup-porting stalks became incorporated ('buried') in the enlarged and confluent laminae of the co-axial sterile pinnules. In the latter case the ovules appear as if borne marginally on a lobe or segment of an assimilatory organ (a 'leaf'!), a condition simulating the traditional postulated 'megasporophyll' (semantically speaking, being *primarily* a reproductive version of a leaf!) but the secondary semophyletic origin of such ovuliferous laminose frond segments rules out their identity with a 'sporophyll' (compare Fig. 7; see MEEUSE 1963b for additional details).

The morphological rigidity of the Cycadofilices, *i.e.*, the retention of the large complex fronds which developed in a rather cumbersome way and could not be rapidly shed to be replaced by new seasonal growth, and of the 'pteridophytic' mode of reproduction (the ovules becoming detached *before* the actual fertilisation, *i.e.*, gametic fusion and embryogeny, took place; see MEEUSE 1963b, CAMP and HUBBARD 1963b), may have caused the rapid decline of these seed ferns after the Carboniferous. They did not evolve into 'advancing' groups, and conceivably their probable dwarf survivors, the Salviniales, only owe their longevity to an early adaptation to an aquatic habitat (MEEUSE 1961b). Among the Cycadofilices two categories of ovules occur, the lyginopterid type (amply discussed by CAMP and HUBBARD) and the neuropteroid (medullosan) *Pachytesta* type. The latter is characterised by a strongly developed outer coat which may have incorporated the cupule and is almost completely fused with the conspicuous inner integument. The presence of a third protective organ of the megasporangium, the cupule, is a characteristic feature of all pteridospermous groups. Its origin from telomes or pre-cauline syntelomes is established by LONG's (1960) work on *Eurystoma*.

When we try to single out the main evolutionary trends among the glossopterid pteridosperms, it appears that, in contradistinction to the Cycadofilices *s.s.*, the complex progymnospermous fronds became generally much more simplified, bipinnate, bifurcate-pinnate, pinnate, digitately compound and, also, that simple fronds (or leaves) are found in groups of glossopterid affinity and in their cycadopsid descendants (Caytoniales, Nilssoniales, Pentoxylales, Cycadales, Bennettitaleans, Chlamydospermae, Angiosperms). Another and considerably more specific development took place in the reproductive regions, the available evidence indicating a spatial segregation of the derivatives of the fertile pinnules of a pinna of an ancestral mixed frond and their co-axial sterile pinnules. The original mixed pinna thus became transformed into a unit consisting of a sterile foliar element bearing on its adaxial side an aggregate of the homologues of fertile pinnules (cupulated megasporangiate or synangial microsporangiate organs). A further segregation, possibly aided by the incorporation of the basal portion of the unit in the cortex of the supporting axis, resulted in a bracteated sporangiate organ,

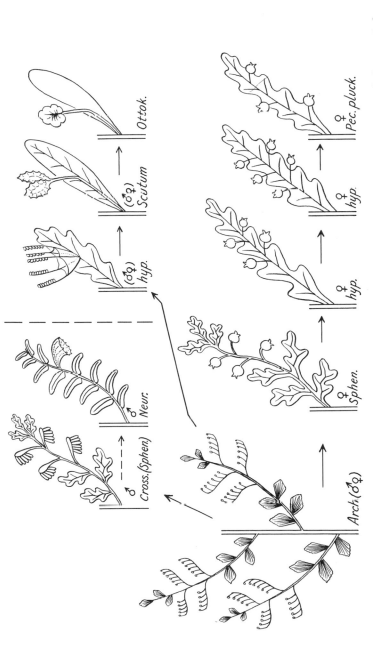

FIGURE 7. Semophyleses of fronds and reproductive regions in Pteridosperms: *Arch.* = *Archaeopteris*; *Sphen.* = *Sphenopteris*; *Cross.* = *Crossotheca*; *Neur.* = *Neuropteris*; *hyp.* = hypothetical stage; *Ottok.* = *Ottokaria*; *Pec. pluck.* = *Pecopteris pluckenetii.*

the gonoclad, which became the fundamental element in the reproductive regions of all cycadopsid groups (see MEEUSE 1963b and also Chapter 14). Since the gonoclad-bract units are the homologues of mixed pinnae of a complex progymnospermous frond, these units often remain co-axially associated in an anthocorm. The cupules, which in the Euramerican Cycadofilices, except the early lyginopterid forms, contain only a single ovule, are usually pluri-ovulate in the glossopterid pteridosperms. This is reflected in the morphology of the cupulated organs of the early Cycadopsids such as the Caytoniales which not infrequently retained the pluri-ovulate condition. The occurrence of bi- to pluri-ovulate derivatives of the pteridospermous cupule among advanced Cycadopsids, even in some angiospermous forms (Amentiferae), must not be precluded.

PROTOCYCADOPSIDA

There is no sharp dividing line between the Protocycadopsida (which include Caytoniales, Corystospermaceae, Peltaspermaceae, Nilssoniales and the recent Cycadales) and their pteridospermous ancestors, but, apart from minor general advancements towards the cycadopsid-angiospermoid type of plant, the complete separation of the gonoclads and their bracts and the attainment of the semophyletic level of the seed may serve as the principal distinguishing character. The always more or less fragmentary remains of the fossil representatives, augmented by the morphology of the Cycadales, give some idea of the characteristics of the group, which must have formed a large and varied assembly, chiefly of Mesozoic age. The Caytoniales with their bilaterally symmetric gynoclads bearing cupulated trusses of ovules and with androsynangia resembling angiospermous staminal thecae, e.g., were clearly different from the Nilssoniales, which bore on their gynoclads biovulate appendages representing reduced cupules and had cone-like androclads with cladodic laminose microsporangiate appendages. Probably owing to the incomplete preservation, fossil forms with complete anthocorms are unknown. The recent Cycads bear solitary gonoclads, but these need not be an indication of the prevailing morphological features of the group as a whole. Their precursors, the Nilssoniales, were probably early specialised (HARRIS 1961, MEEUSE 1963a). At any rate, there is irrefutable evidence that related groups of Mesozoic age such as the Pentoxylales bore more complicated reproductive structures which are evidently anthocorms, and in any event the only possible pathway for further evolutionary advances of the floral region culminating in the angiospermous 'flower' requires ancestral forms with anthocorms, so that one must postulate that some or the majority of the fossil Protocycadopsids bore their gonoclad in groups co-axially arranged about a common supporting axis. The chances of the more or less complete fossilisation of such complicated structures are of course rather small, but a lucky find may settle the point.

HIGHER CYCADOPSIDS

The Higher Cycadopsids constitute a fairly heterogeneous assembly including, apart from the Angiosperms, various Bennettitalean (cycadeoid) groups, the Pentoxylales, and taxa of chlamydospermous affinity. It is among some of these forms that angiospermous features developed and it is, therefore, necessary to assume that at least some of them had primitive anthocorms. Such complex reproductive structures occurred in the Pentoxylales and that is one of the reasons why it is feasible that pentoxylalean archetypes gave rise to Monocotyledons (MEEUSE 1961a). However, the 'strobilus' of *Cycadeoidea* is anything but a clear-cut case. Although, according to DELEVORYAS (1963), the Bennettitalean strobilus was functionally unisexual ('protandric'), it is morphologically amphisporangiate. Considering that ovules and male synangia are comparable morphological entities (they are both homologues of progymnospermous fertile pinnules!), at first sight the most likely interpretation seems to be that the strobiloid reproductive organ is an amphisporangiate gonoclad, female in the distal and male in the proximal region, because the sporangiate appendages are bractless (the basal bracts forming a kind of 'perianth' are anatomically not associated with the microsporangiate appendages!). Such a 'strobilus' and the corresponding unisexual organs of the Williamsonieae could then be interpreted as structures comparable with the cone-like gonoclads of the Cycadales. This interpretation of the strobile of the Cycadeoidales denies the homology of these organs with a magnoliaceous (Ranalian) flower, which is, in my opinion, a modified ambisexual anthocorm bearing numerous gonoclads. On the other hand, the complicated male reproductive structures originally bearing a large number of synangia and the numerous cupulated ovules are also suggestive of a more complex structural pattern of the microsporangiate lateral and the ovuliferous terminal portions of the Bennettitalean strobilus. The conditions in the reproductive regions of the more or less closely related chlamydosperms do not rule out the possibility of the cycadeoid strobilus being, after all, a modified anthocorm with a single terminal (or secondarily pseudo-terminal) gynoclad and a number of lateral androclads which lost their bracts. The sterile proximal elements forming the 'perianth' would then represent the sterile phyllomes commonly occurring in the basal region of an anthocorm. This would nevertheless also rule out a direct connection between the Cycadoideales and the Polycarpicae, because according to this interpretation the bracts of the androclads are wanting in the first category and the solitary terminal gynoclad of these Bennettitalean forms does not correspond with the numerous lateral gynoclads of the Ranalian flower. The Cycadeoideales are undoubtedly specialised offshoots of a cycadopsid lineage and also too recent to qualify as angiosperm ancestors because several of them are contemporaneous with the early (Cretaceous) Angiosperms.

The morphology of the reproductive regions of the Chlamydosperms is obscured by the extreme oligomerisations and reductions in their anthocorms. The ovules, discussed in Chapter 15, are undoubtedly hemi-angiospermous, and other features such as embryological, palynological and anatomical characters point in the same direction. The morphologically but not always functionally bisexual fertile regions are also suggestive of angiospermous trends, whilst providing some possible clues concerning the origin of the bisexual anthocorm (later becoming the bisexual 'flower'). The problem of the distribution of the sexes cannot be solved by morphological inquiries alone, because physiological and ecological factors must have played an important part in the selective (adaptive) evolution of the floral region which is undoubtedly associated with the mutual relationships between entomophilous Protangiosperms and anthophilous insects. From a purely morphological point of view, the apparently predominantly unisexual mixed fronds of the Progymnosperms, the unisexual protocones and cones of the Coniferophytina, the unisexual gonoclads of the Protocycadopsids, and the unisexual anthocorms of the Pentoxylales suggest that primarily, at least, the gonoclads were almost consistently unisexual (exceptions are the Marsileales with amphisporangiate cupules, perhaps Glossopteridales and *Vojnovskya*). All that is needed to produce a bisexual entity is the coaxial arrangement of male and female gonoclads on a common supporting axis to form an amphisporangiate anthocorm, a condition that could easily develop when changes occur in the physiological (biochemical) processes underlying sex determination. Partial conversions of unisexual reproductive structures into amphisporangiate complexes are known to occur in Conifers (androgynous cones) and Angiosperms in a wide range of cases varying from occasional teratological developments to rather constant and genetically controlled variations. It only requires some selective pressure to increase the relative frequency of occurrence of these 'hopeful monsters' to arrive at bisexual anthocorms. This implies that the angiospermoid Bennettitalean-chlamydospermous Cycadopsids were already diversified in this respect, so that consistently diclinous (monoecious or dioecious) and more or less completely monoclinous (bisexual) forms occurred side by side (HESLOP-HARRISON 1958). These primary monoclinous and diclinous conditions were simply inherited by the angiospermous groups polyrheithrically evolved from protangiospermous stock, so that some groups, such as the Monochlamydeae, Pandanales and Arecales, are still predominantly diclinous and other ones, such as the Polycarpicae and the Liliales, mostly ambisexual. The Chlamydospermae provide a clue, already recognised by, *e.g.*, PEARSON (1926), in that sterile ovulate structures occur in the male reproductive region. The first pollinating insects were either attracted by the pollen as a source of food (beetles with biting mouth parts), or by the pollination drop exuded by the elongated tubillus (micropylar tube) of the inner integument (insects with licking or sucking mouth parts, such as flies, Hemiptera and Hymenoptera). For a successful pollination, both male and female reproductive structures must be visited by the same insects, but if an insect either collects only pollen of the male genitalia, or visits only the ovuliferous organs, only chance pollinations could possibly result, unless the male and the female genitalia are borne in close proximity. The male *Welwitschia* plants attract insects which feed on the liquid produced by the sterile ovules topping the functionally male fertile structures. Significantly, the micropylar tube is distally enlarged into a peltate disc on which the nectar collects. The microsporangia of the male genitalia closely surround the disc and their pollen is thus easily carried away, eventually to be deposited on the fertile ovules of a female plant. The conditions in *Gnetum* and in some species of *Ephedra* are rather similar, sterile or occasionally even fertile ovules occurring in the distal portions of the male reproductive region.

The most likely morphological interpretation of the gnetalean fertile region, in the light of the above-mentioned considerations, is that they represent bisexual or unisexual anthocorms. However, these anthocorms, like the flowers of their probable nearest angiospermous allies, the Piperales, have become very much reduced by the progressive oligomerisation of the number of gonoclads and of the number of chlamydote ovules or stalked male synangia per gonoclad. The fairly complicated reproductive structures of *Welwitschia* and *Gnetum* reflect their origin from compound anthocorms. The male 'cone' of *Welwitschia*, for instance, could be interpreted as consisting of a number of co-axial bracteated reduced ambisexual anthocorms, each containing two bracteated androclads reduced to a few 'stamens', and a terminal gynoclad reduced to a single ovule. However, this is not the only possible interpretation. There is some evidence of an incipient monocliny among the recent Chlamydospermae which may have been brought about by the advent of amphisporangiate gonoclads (androgynoclads). There is a moot point, especially in such reduced reproductive regions as those of the contemporary gnetalean forms (see also MELVILLE 1963). Several angiospermous lineages must have descended from forms with chlamydospermous characters but with more primitive (*i.e.*, unreduced) anthocorms bearing numerous gonoclads, each with several ovules or several male synangia (stamens). A characteristic and apparently fairly constant feature of the Higher Cycadopsida is the mono-ovulate cupule, which differentiated in several directions and appears as the fleshy 'chlamys' in *Gnetum* and *Carnoconites* (Pentoxyales) or as the interovular scales

PLATE II. A cauliflorous and apparently scandent *Gnetum* species, photographed by Dr. W. MEYER in a primaeval forest, Nunukan, Borneo. The spikelike ovuliferous structures borne on the older parts of the stem. (Courtesy of Flora Malesiana Foundation.)

in the Cycadeoideales. The 'disintegration' of the cupule into a number of elements (*e.g.,* interovular scales) may be the explanation of the apparent absence of a third ovular coat in *Ephedra* and *Welwitschia* (and perhaps in some angiospermous taxa such as Myricaceae). If one assumes that the ovules of such taxa are secondarily unitegmic by the loss of the outer integument, the outer ovular envelope is the cupule (chlamys); if they are considered bitegmic, the cupule would have disappeared. A third and to my mind more plausible interpretation is that the chlamydospermous ovule is essentially bitegmic and that the cupule is sometimes represented by the elements forming what is conventionally called the 'perianth' or the 'prophylls' of the 'female flower' of *Welwitschia* and *Ephedra* (see also MEEUSE 1964b). The morphological diversity of the higher Cycadopsids, in conjunction with the unmistakable angiosperm tendencies in their embryological, anatomical, palynological and anthecological features, strongly suggest that there was not one clearly defined group of hemi-angiospermous progenitors from which all recent Flowering Plants descended but rather a number of more or less coetaneous taxa, each of which gave rise to a group of Angiosperms (*e.g.*, pentoxylalean forms to Monocots, chlamydospermous archetypes to such groups as Piperales, and various less specialised forms of gnetalean and cycadeoid affinity to other groups of Dicots such as Polycarpicae, Dilleniales and Monochlamydeae).

A fundamental consideration is the lack of cogent indications of phyllospory among the protocycadopsid and chlamydospermoid-Bennett-italean groups. The so-called megasporophyll of *Cycas* is incongruous and almost certainly of secondary origin (MEEUSE 1963a). Even the equally famous 'microsporophylls' of *Cycadeoidea*, since WIELAND's reconstruction universally reputed to be frond-like expanded organs bearing marginal microsporangia until DELEVORYAS (1963) re-examined them carefully and came to an altogether different conclusion, are not unequivocally conformable to a postulated 'sporophyll'. The interpretation of the reproductive organs of the Angiosperms must be based on the morphology of the Higher Cycadopsids and phylogenetically deduced in terms of the continuous descent of the Flowering Plants from Devonian Progymnosperms through pteridospermous, protocycadopsid and cycadopsid levels of organisation. A quest for Protangiosperms starting from 'reconstructed' archetypes and clouded in classical Angiosperm-centred notions was doomed to futility. The phylogenetic approach of the interpretative morphology of the floral region is probably the most convincing demonstration of the inadequacy of the conventional morphological dicta.

12

Phylogeny of the Reproductive Region: III. The Various Aspects of Angiospermy

The Traditional 'Unbridgeable' Gap Between Gymnosperms and Angiosperms. Suggested Origins of Angiosperms. Criteria of Angiospermy. Continuity of Cycadopsids and Angiosperms. 'Age' of the Flowering Plants.

A little over a hundred years ago, phytomorphology began to grasp the fundamental differences between the Angiosperms and those recent groups of seed-bearing plants which had previously been associated with the Dicotyledons but constitute what we now call the Gymnosperms. Palaeobotany still being in its infancy, it was on the ground of typological considerations based on the postulates of the idealistic morphology that the botanists concluded that the reproductive organs of the Coniferae and other plants did not fit the pattern of an angiospermous 'flower' and are apparently based on a different archetype. The comparative morphology of the reproductory processes, so brilliantly conceived by HOFMEISTER, indicated a homology between the ovules of the seed-bearing Cormophyta, but the classical concept of the 'carpel' inevitably led to the conclusion that the ovules are enclosed in an ovary in all true Flowering Plants, so that the groups with exposed ('naked') ovules could not be included in the Angiosperms any longer. This classical postulate thus provided the conventional distinguishing character between Gymnosperms and Angiosperms, and also their respective names, the ovules of the Gymnosperms being 'naked', *i.e.*, not encased in a 'sporophyll', and those of the Angiosperms protected or covered by the enveloping 'carpel', which formed a containing 'vessel'.

After contemporaneous gymnospermous plants had been studied more extensively and several fossil groups had been discovered, the comparative morphology and anatomy pointed to relationships between Gymnosperms and Angiosperms, so much so that by the turn of the century two major working hypotheses regarding the phylogenetic origin of the Angiosperms emerged which are both based on the assumption that the Flowering Plants have descended from gymnospermous ancestors. These theories are generally known as the *euanthium theory* (anthostrobilus theory) of ARBER and PARKIN (1908) and the *pseudanthium theory* of DELPINO-WETTSTEIN (see, *e.g.*, R. VON WETTSTEIN 1935). The first was founded on a hypothetical group of Protangiosperms, combining characters of Cycadeoidales (the 'strobilus', supposed to be functionally bisexual, *i.e.*, monoclinous, and the male reproductive organs, the so-called microsporophylls), of *Cycas* (the female reproductive organ, the 'megasporophyll'), and of the Magnoliales (various characters supposed to be primitive: wood anatomy, type of leaf, etc.).

The 'gap' between the Angiosperms and the Gymnosperms was thought to have been bridged by unknown extinct forms with 'open' laminose sporophylls of which the female ones became follicular and encased the ovules. The two basic assumptions, *viz.*, the origin of the Flowering Plants from the 'reconstructed' preconceived forms with primitive 'sporophylls' ('Proranales', etc.) and the status of the Magnoliales, or at least the Polycarpicae, as the most original type of Angiosperm, still prevail in the majority of the taxonomic and morphological publications (see, *e.g.*, HUTCHINSON 1959, TAKHTAJAN 1959b). The main difficulty was the lack of suitable fossil forms which might come up to the requirements of the hypothetical Protangiosperms of ARBER and PARKIN, apart from the growing realisation that anatomically (BAILEY and his school: see, *e.g.*, BAILEY 1954) and palynologically the 'Polycarpicae' are a heterogeneous assembly. However, this did not make the botanists aware of possible flaws in their premises; on the contrary, the euanthium theory became to be regarded as factual, and various suggestions were made to explain the lack of palaeobotanic records of prospective ancestral forms that would substantiate the theory. I have discussed this point elsewhere and explained that the theory is based on some erroneous assumptions and false postulates, so that it is untenable. The search for the reconstructed 'primitive Angiosperms' was doomed to remain unsuccessful, because they have never existed, and all explanations to account for the apparent absence of pre-Cretaceous fossil progenitors of the Angiosperms are entirely superfluous (MEEUSE 1962). I am of the opinion that the remains of several ancestral taxa of the Flowering Plants have already been discovered, but the traditional and ingrained ideas did not help to create a receptive state of mind, so that these fossils remained unrecognised because they did not answer to the description that had been broadcast.

The alternative pseudanthium hypothesis is also an attempt to link the Angiosperms with gymnospermous taxa, the supposed connecting group being sought among chlamydospermous forms which are thought to lead, through such forms as *Casuarina*, in the first place to the Amentiflorae. The phylogenetic prototypes are more or less tangible because they were assumed to have had much in common with the recent Chlamydosperms, and the genitalia at any rate were supposed to be very similar to those of the concrete examples of living forms such as *Ephedra* and *Gnetum*. The principal stumbling block of the theory is the required derivation of a closed, pluri-ovulate and apocarpous Ranalian carpel, with laminal or submarginal placentation, from the solitary terminal ovule (or from several such ovules) of the postulated prototype (the primitive pseudanthium), especially because the reputedly ubiquitous 'sporophyll' had to be conjured up from somewhere. NEU-MAYER (1924) came very near the solution by substituting fertile *axes* for the gnetalean 'female flowers' (ovules!) and 'male flowers' (stamens) of the original version of the pseudanthium theory, but he retained the solitary ovule on each ovuliferous axis (gynoclad) and stranded on some highly fanciful additional assumptions which lead, among other things, to utterly unacceptable derivations of the apocarpous gynoecia of the Polycarpicae. The fact that, in the pseudanthium theory, the carpels are inescapably of secondary origin and are at best only pseudo-carpels derived from non-ovuliferous (sterile) bracts which became associated with the ovules to form the closed ovaries of the Angiosperms, was not clearly understood even by the majority of the protagonists of the idea of the polyaxial flower (except NEUMAYER), but was recognised by, *e.g.* HAGERUP, FAGERLIND and LAM. Others make very light of this difficulty because they somehow manage to find sporophylls in the angiospermous pistil, irrespective of its phylogenetic history. JANCHEN (1950), a declared adherent of WETTSTEIN's views, is so muddled by traditional morphological concepts and inaccurate definitions that he rejects LAM's 'stachyospory' altogether and firmly maintains the appendicular interpretation of the genitalia—a good example of how Angiosperm-centred connotations of the conventional phytomorphological fundamentals (postulates), definitions and semantics lead to muddled thinking.

PANKOW (1962), although a firm believer in 'stachyospory' and 'phyllospory' as alternative conditions in the gynoecia of the Angiosperms, does not distinguish 'true' carpels (*i.e.*, derivatives of the postulated sporophylls) and pseudo-carpels (*i.e.*, sterile elements secondarily associated with ovuliferous axes), because he does not think such a distinction has much practical significance, but this is a denial of the fundamental character of the alternative conditions emphasised by LAM. LAM's concepts of 'stachyospory' and 'phyllospory' (1948 *et seq.*) go back to an idea expressed by SAHNI as early as 1921. The rather diverse morphology of the Gymnosperms, especially of their reproductive organs, caused SAHNI to believe that there are at least two main groups of the Gymnosperms, one with axis-borne seeds (the Stachyospermae) and one with leaf-borne seeds (the Phyllospermae). LAM extended this classification to all Cormophyta and changed the names to 'Stachyosporae' and 'Phyllosporae' to include the sporangiate (reproductive) organs of both sexes. He considered the two conditions of stachyospory and phyllospory to be so fundamental that he even split up the Angiosperms in two groups on the ground of this difference, which he supposed to reflect the phylogenetic origin of the genitalia. The Flowering Plants are, in his opinion, heterogeneous and must have had at least a dual origin (dirheithry or biphyly), some of them having descended from stachyosporous ancestors and the remainder from phyllosporous forms.

I do not subscribe to this supposed fundamental difference, as I have explained in another chapter, because I think that all Spermatophyta have had a common group of progenitors (BECK's Progymnosperms) which, on the ground of the morphology of their reproductive organs, can only be interpreted as 'stachyosporous' (*i.e.*, with axis-borne sporangia). The Progymnosperms soon evolved along three main lineages, one leading to the Coniferophytina, and another to the seed ferns (*sensu lato*), the latter lineage separating again into the evolutionary lines of the true Cycadofilicales of the Euramerican coal seams and of the glossopteridalean seed ferns of the ancient Gondwana region, of which only the latter differentiated appreciably and produced, through Protocycadopsida and Cycadopsida, the main lineage leading to the Angiosperms. The often considerable differences between the various groups of the Gymnosperms (all initially descendants of the Progymnosperms) are the result of the long phylogenetic history of each major group and of the prevalence of certain more or less specific evolutionary trends in each main phylogenetic line, but there is a fundamental homology resting upon their common descent from corresponding archetypes. It is my firm conviction that the condition of 'angiospermy' must have developed as a trend in the cycadopsid Gymnosperms and so far the evidence seems to be compatible with this working hypothesis. If the phylogenetic relationship between cycadopsid groups and the Flowering Plants is accepted, the tangible advantages of this simple assumption become evident, provided the Angiosperms—the youngest and most advanced forms!—are no longer taken as the starting point of phylogenetic speculations. Quite simply, the inquiry resolves itself into a selection of those features which are more or less exclusively confined to the Flowering Plants and thus 'define' them as Angiosperms whilst distinguishing them from the majority of the cycadopsid Gymnosperms, followed by the interpretation of these specific characters in terms of the morphology of the gymnospermous groups instead of fitting hypothetical groups into the conventional angiospermous pattern.

There are anatomical, palynological, embryological and some miscellaneous data relevant to this problem. As regards the anatomy, the presence of xylem vessels is one of the characters said to be rather exclusive with the Angiosperms, but this is an exaggerated claim which is also inaccurate in that only the occurrence of vessels in the *secondary* xylem, a derivative of the lateral meristem called the (stem) cambium, is significant. All Dicotyledons, except the well-known vessel-less genera mainly belonging to the Polycarpicae (and some highly specialised aquatics and parasites which can be disregarded in this connection), do indeed possess secondary xylem vessels, and this is in striking contrast with the Gymnosperms (the Chlamydospermae excepted). There is no convincing evidence that the Monocots descended from forms with the dicotyledonous type of secondary growth. The Monocots have a fundamentally different stelar anatomy (matched only by that of some Piperales and Nymphaeales), and the structure of the stems of the forms exhibiting secondary growth can be compared with that of the gymnospermous Pentoxylales (MEEUSE 1961a). Monocotyledons with secondary growth (*Dracaena, Cordyline, Yucca, Aloë,* etc.) produce secondary xylem which consists exclusively of fibre tracheids (usually with bordered pits), and the occurrence of vessels in the primary xylem is irrelevant in this connection. CHEADLE (*e.g.,* 1953) has contended in a series of papers that the vessels in Monocots are most probably of independent origin and not directly semophyletically related to the vessels of the Dicots. There is, therefore, no cogent reason to accept a monophyletic origin of the dicotyledonous xylem vessels. After all, vessels developed independently, undoubtedly as a functional adaptation, in such unrelated groups as ferns (*Pteridium*), Lycophyta (*Selaginella*) and various cycadopsid groups.

The absence of vessels in the secondary xylem is a primitive feature, and the finds of certain pre-Cretaceous 'homoxylous' woods do not materially contribute to the solution of the problem of the origin of the Angiosperms—these 'homoxylous' fossil woods may equally well represent the remains of the stems of Bennettitalean forms as those of early dicotyledonous Angiosperms (which amounts, presumably, to the same thing, but this is a conclusion deduced by inference, not from the structure of the fossil stems!).

The outcome of these considerations is clearly that the presence of vessels is a rather poor criterion by which to define an Angiosperm.

The stomatal apparatus is rather diverse among the Angiosperms; the main types can be interpreted as conceivably derived from either a syndetocheilous or a haplocheilous ancestral form, so that one cannot define an angiospermous condition by the morphology of the stomata alone.

The occurrence of Jurassic 'dicotyledonoid' sporomorphs, indeed referred to dicotyledonous Angiosperms by some authors, such as PFLUG (1953), but attributed to unspecified 'gymnosperms' by other workers (see, e.g., HUGHES 1961b), is to my mind a clear indication of the appearance of prototypes of dicotyledonous pollen forms. The occurrence of germination slits or pores as a requisite for angiospermous pollen does not stand in the way of its *origin* from similar but still 'gymnospermous' sporomorphs. HUGHES (1961a) has found that the form genus *Eucommiidites* is associated with ovules suggesting a chlamydospermous affinity. In other chapters I shall indicate the close relationships between chlamydospermous Cycadopsids and some dicotyledonous taxa, so that at least some of the pre-Cretaceous angiospermoid pollen types must definitely have belonged to protangiospermous groups or perhaps already to angiospermous plants; this is largely a matter of opinion, as it depends on the circumscription of an 'Angiosperm', as I shall presently point out.

The embryological features of truly angiospermous forms include the reduction of the gametophytes, the consistent occurrence of siphonogamy with double fertilisation, often followed by the formation of a secondary endosperm, and the complete development of the embryo. I have shown in detail elsewhere (MEEUSE 1964a) that the process of double fertilisation is an almost inevitable consequence of the perfection of the process of siphonogamy in conjunction with the development of 'angiospermy', and that conditions resembling double fertilisation already occur in the Chlamydospermae. Other evidence (GERASSIMOVA-NAVASHINA 1961) supports the conclusion that double fertilisation is neither a singular phenomenon that originated only once during the evolution of the Angiosperms (the traditional argument in favour of their monophyletic origin!) nor an exclusive feature restricted to the Flowering Plants. The advent of the double fertilisation process is much older than the attainment of the complete level of angiospermy, of which it is only one aspect. The production of a secondary endosperm is not a constant feature and therefore inadequate as a specific character to distinguish *all* Angiosperms from other spermatophytic groups.

In cycadopsid Gymnosperms, the embryo usually attains its maximum development only after the seed has been shed and matures when the detached seed is lying on the ground or in the soil, after which germination follows immediately, so that there is no resting stage (Cycadales, *Gnetum*). In typical Angiosperms, the seed becomes detached from the mother plant only when the development of the embryo is completed and it has a resting stage between the termination of the growth of the embryo and the beginning of germination, but there are some exceptions.

The seeds of Chloranthaceae fall off before the embryo is fully developed (YOSHIDA 1959).

All the characters considered so far being neither exclusive nor always constant, they are insufficient to distinguish every traditionally angio-spermous taxon sharply from the Gymnosperms. We are thus left with the original differential (diagnostic) character, the condition of 'angio-spermy' ('angiovuly', protection of the ovules, is a term preferred by some authors) and this criterion seems to be the crucial one. Not only phytomorphologists, but also students of 'floral biology', of conditions and relations associated with the process of pollination, have always main-tained that there is an 'unbridgeable gap' between the functional gynoecia of the Gymnosperms, in which the ovules or their individual protective layers (such as the integuments) perform the pollen-receiving function themselves, and those of the Angiosperms, in which the recep-tion of the pollen occurs on the stigmatic portions of the 'carpels'. The closing of the gap was said by VAN DER PIJL (1961) to require some mysterious transference of the function from the ovule to the ovary wall, so that any suggested relation between, *e.g.*, *Gnetum* and the Angio-sperms was rejected beforehand as an impossibility.

This is another example of botanists being blinded to the recognition of certain phylogenetic connections by traditional concepts. In Chapter 15, I shall discuss the semophylesis of the ovule and demonstrate that there are cogent arguments to adopt an alternative interpretation of the conventional single-ovuled 'ovaries' (pistils) of many Monochlamydeae, Piperales and Spadiciflorae, based on the assumption that there is no such 'gap' but a direct phylogenetic connection between the cycadopsid Gymnosperms with chlamydote ovules and these angiospermous groups. This implies the homology of the mono-ovulate female genitalia, tradi-tionally pseudo-monomerous carpellate organs, of certain angiospermous orders with chlamydote gnetalean ovules, *i.e.*, with derivatives of pterido-spermous cupules. The two conditions, *viz.*, the gymnospermous ovule with exposed micropyle and the closed 'angiospermous' gynoecium with hidden micropyle, are semophyletically continuous, not, as I have previously (1963) suggested, because conceivably the chlamys (aril) or outer integument (or both) gradually overgrew the exposed micropyle of the inner integument and a transference of the pollen-receiving func-tion from the exposed exostomium of the micropylar tube of the inner integument to a part of the closing inner integument or the overarching chlamys took place. Intermediate semophyletic stages exhibited by *Engelhardia spicata* and *Canacomyrica* rather suggest the retention of the original exposed exostomium and of the stigmatic areas which catch the pollen grains, a condition which is actually only pseudo-angio-spermous (MEEUSE 1964b, MEEUSE and HOUTHUESEN 1964). A gradual transference of function took place during the semophylesis of the closed gynoecia of the carpellate type, which developed from an exposed ovuliferous axis and its bract (or from several such axes and associated bracts; see Chapters 14 and 16). The ovules must originally have caught the pollen individually on their micropyles, but when the ovaries gradually became completely closed structures the pollen grains could not become introduced into the interior cavity of the gynoecium any more, so that they germinated on the outside and sent their pollen tubes from the stigmatic areas down into the ovary chamber. In the gynoecia of *Degeneria* and of *Drimys* section *Tassmania* a transitional phase is still extant: the slit in their incompletely closed carpels is covered by matted hairs which catch the pollen, but do not prevent the pollen tubes from penetrating through the meshes between the hairs; these tubes have actually still free access to the ovules. The closure of the slit by carpellary tissue resulting in the formation of a 'stigmatic crest' completes the transition from 'gymnospermy' to 'angiospermy' in these forms.

Upon closer examination, *all* the accepted points of difference between Gymnosperms and Angiosperms break down and a gradual passing of conditions prevailing among the cycadopsid Gymnosperms into the characteristic features of the Flowering Plants seems to be the rule. The Angiosperms have thus been relegated from their position as an isolated group of dubious origin to the modest but much more convenient status of specialised cycadopsid Gymnosperms. The definition of 'an Angiosperm' becomes a matter of opinion, because there is no sharp dividing line between them and Bennettitalean-chlamydospermous Cycadopsida. The prominent characteristics of the Flowering Plants must be seen as the ultimate phases of semophyletic developments, including anatomical, palynological, embryological and ecological evolutionary tendencies, which can be briefly referred to as 'Angiosperm trends'. It is tempting to ascribe the spectacular and almost precipitous ascent of the Flowering Plants to the amazing reciprocal adaptation between the higher plants and pollinating insects (*e.g.,* GRANT 1950, LEPPIK 1955–60). There can be very little doubt that such specialised families as Labiatae and Orchidaceae have been 'made' by advanced groups of the Hymenoptera, but it is the *early* advent of the plant-insect relationship which is important. Various students of floral biology accept an evolution of insect flower types from primitive 'beetle flowers' to advanced entomophilous, ornithophilous, chiropterophilous and other categories, but I think this is far too schematic, and also suggestive of an evolution of *all* 'flowers' from the same Ranalian (magnoliaceous or nymphaeaceous) archetype. I cannot help feeling that an interpretation dovetailing so nicely with the morphological euanthium theory of the origin of the angiosperm 'flower' was largely inspired by the prevailing ideas regarding the primitive flower and the position of the Polycarpicae in pseudo-phylogenetic systems. The pollination of the peculiar inflorescences of the undoubtedly ancient genus *Ficus* is accomplished by small Hymenoptera. The interdependence between the figs and the gall wasps, combined with the enormous speciation of the insects and their respective hosts that resulted in a high specificity of the wasp-fig relation (CORNER, VAN DER VEGT; see WIEBES 1963) can be reconciled only with a very long evolutionary history of this mode of pollination. It is most unlikely that the moraceous inflorescence developed from large Ranalian 'beetle flowers' with numerous perianth lobes, and the assumption that the Moraceae descended from Ranalian or 'pro-Ranalian' forms is equally untenable.

If, like the present writer, one accepts the view that 'Angiosperm trends' developed in several already independent but still 'gymnospermous' lines of descent, a multiple descent or 'polyrheithry' must be postulated and the traditional class of the Angiosperms must be split up in such a way that the recent taxa belonging to a single evolutionary line based on the same ancestral cycadopsid group are united. The lineage of the Monocotyledons, for instance, may have had a long individual existence (MEEUSE 1961a) and attained the level of 'angiospermy' independently of the Dicots. The more primitive types of monocotyledonous plants (Pandanales, Arecales) do not bear reproductive structures resembling the Ranalian 'beetle flowers', but the large and sweet-scented male 'inflorescences' of *Pandanus* and other forms are nevertheless visited by pollen-collecting insects, including bees and beetles. As far as the morphological consequences of the development of various types of insect-pollinated reproductive organs are concerned, one must anticipate a great deal of parallelism. This is very important for the interpretation of the evolution of the Angiosperms, which is based mainly on the semophylesis of their reproductive region. A correspondence in morphological characters between two angiospermous taxa, such as the typological or 'phenetic' resemblance between the flowers and genitalia of more advanced Monocots (*e.g.*, Liliales) and those of dialypetalous Dicots, has rather generally been accepted as an indication of a direct semophyletic connection, but the 'showy' flower type of the predominantly entomophilous groups of the Monocots such as Liliales and Orchidales developed independently of the flowers of the Dicots. Semophyletic trends in specialised monocotyledonous families which developed, for instance, sympetaly and zygomorphy (Orchidales!) must be the result of parallel adaptive evolution independently of the advent of analogous 'Gestalt' types in representatives of the Dicots such as Labiatae and other 'Personatae'. I cannot accept the 'euanthous' Ranalian beetle flower as the one and only archetype of the angiospermous 'flowers' (an idea strongly advocated by Leppik), certainly not for the Monocots, the Monochlamydeae and the Piperales, and only with a question mark for the Centrospermae and the Dialypetalae, which leaves but little scope. Incidentally, not all groups constituting the heterogeneous Polycarpicae

PLATE III. *Pandanus tectorius,* growing in quantity on coral limestone, along the coast of Nusa Barung (southeastern Java). (Photo by M. Jacobs, May 1957. Courtesy of Flora Malesiana Foundation.) Compare Plate IV.

bear more or less typical euanthous flowers: they occur, for instance, among the Magnoliales, Nymphaeales, Nelumbonales and Ranunculales, but not in Laurales, Trochodendrales, and Hamamelidales.

The likelihood of a polyrheithric descent of the Angiosperms and the frequent occurrence of homoplasies must put us on guard against all generalisations based on typological considerations, such as the qualification of certain features as 'primitive' or 'advanced'. The same 'character' may be more primitive, *i.e.*, phylogenetically older, in one lineage and much more 'derived' in a different group. Carpellate ovaries are, in my opinion, always secondarily evolved from associations of ovuliferous axes (gynoclads) with their subtending bracts, but closed ovaries did not develop simultaneously in the individual evolutionary lines, nor did they always become a universal feature in every group. Sometimes an association of one gynoclad and its bract predominated (Magnoliales *s.s.*, Ranunculales), but in other taxa a number of gynoclads and their bracts developed simultaneously into a coenocarpous ovary (Guttiferae, Juncaceae, many Liliiflorae), and this led to the development of *alternative* categories which are not derived from each other but are related only through a common archetype which had a rather different morphology. One cannot maintain that the ovary of the Guttiferae is more advanced than that of the Magnoliaceae, let alone that the coenocarpous gynoecia of the former originated from the fusion of apocarpous Ranalian follicles. On the other hand, the more advanced condition of the gynoecia of *Freycinetia*, Juncaceae and Liliales in respect to those of *Pandanus*, of the Saururaceae in respect of those of the other Piperales, and of Salicaceae and Fagaceae in respect of those of Juglandaceae can presumably be accepted. There are more examples of alternative characters, of which the arborescent and herbaceous growth forms provide a good example. Woody and herbaceous forms are sometimes phylogenetically related, herbaceous forms being derived from woody ancestors, or *vice versa*. Most authorities accept a relation between Capparidaceae and Cruciferae, Bignoniaceae and Pedaliaceae-Acanthaceae-Scrophulariaceae, Araliaceae and Umbelliferae, Pandanaceae and Typhaceae-Sparganiaceae, respectively, usually the herbaceous taxa—correctly in my opinion—being regarded as representing the more derived forms. The cases mentioned are examples of recurrent parallelisms, and it is one of the weaknesses of HUTCHINSON's classification that far too much significance is attributed to the habit, so that in his system, for instance, Capparidaceae are far removed from Cruciferae, and Bignoniaceae from Pedaliaceae and Acanthaceae. In other groups, the herbaceous or woody habit is a retained primary condition. I do not believe that, for instance, *Nelumbo* and Nymphaeaceae descended from caulescent Cycadopsids, or the Amentiflorae from small herbaceous forms, and this provides another example of the occurrence of alternative conditions (a geophytic and an arborescent habit, respectively) in parallel phylogenetic lines. Similar alternative growth forms occur in the Monocots: the arborescent Pandanaceae and Palms opposite the rhizomatous Typhaceae, Sparganiaceae, Cyperaceae and Cyclanthaceae; the caulescent genus *Dracaena* and the related geophytic *Sansevieria*.

Another example of alternative characters is the occurrence of 'bisexual flowers' and 'unisexual flowers', some taxa being originally 'hermaphrodite' (Dicots generally, except the Monochlamydeae, the majority of the Liliiflorae, Scitamineae, etc.), and other ones predominantly diclinous (Monochlamydeae, Pandanaceae, Restionaceae, Araceae, Arecaceae, etc.).

The assessment of 'primitive' and 'derived' stages of characters to obtain a 'relative' advancement index (SPORNE 1956) is virtually useless because the pooling' of the various terminal taxa of the parallel lineages, *i.e.*, of homologous, homoplastic and alternative conditions, to judge them by the *same* standard (which can only apply to one or a few of such lines but not to all of them) is not likely to yield any reliable information about the true phylogenetic relationships. Parallelisms (convergencies) and alternative conditions cannot be compared in terms of an advancement index presupposing a linear semophyletic relationship of primitive and advanced stages of strictly *homologous* organs.

This is also a warning against overrating the methods of classification by means of calculating devices ('data processing') which is so much *en vogue*. It is simple enough to feed appropriately coded information into an electronic computer to obtain an answer but it is not such an objective method of establishing taxonomic relationships as some systematists are inclined to believe. The coding of morphological taxonomic features is the moot point, because it is (and probably will remain) a matter of opinion whether typologically very similar characters in two different taxa are to be assessed as homologues belonging to the same class of related (comparable) entities, or as analogues, parallelisms, or alternative conditions representing more or less unrelated to fairly intimately connected elements. The subjective judgment has not been eliminated as long as such differences are not satisfactorily accounted for, and only the results of multiple correlations based on most judiciously coded data can be accepted as *possibly* significant.

The Angiosperms, as we have seen, are much more heterogeneous than is generally accepted and represent the terminal taxa of several parallel phylogenetic lines all rooting in gymnospermous groups. Even the conventional 'flowers' are not necessarily all morphological equivalents (see Chapter 18). It thus becomes increasingly difficult to define an Angiosperm, and especially to draw the line between certain groups of Flowering Plants and Cycadopsids of Bennettitalean or chlamydospermous affinity, because all the traditional distinguishing features break down. This is only to be expected when one is searching for direct phylogenetic relationships; and all one can say is that in several parallel lineages a number of Angiosperm trends developed, so that any taxon in which all Angiosperm trends reached the highest level of evolution is indubitably an Angiosperm. However, there are some groups in which one or several trends have not reached the ultimate level (the Chloranthaceae being a case in point), and any decision regarding the status of such forms is inevitably an arbitrary one. This has considerable bearing on another much-debated subject, the question of the 'age' of the Angiosperms (e.g., AXELROD, 1950, 1961, HUGHES 1961b, SCOTT et al. 1960). The combatants in this paper war cannot have had a very clear picture of the stakes they were fighting for. AXELROD has attempted to make out a case for an early origin of the Angiosperms, his opponents maintaining that they are much younger, but neither side clearly defines what an 'Angiosperm' is. HUGHES denies the occurrence of unquestionably angiospermous pollen before the Cretaceous, but this implies a definition of an Angiosperm as a plant with a special kind of sporomorph, which is a rather one-sided point of view. The angiospermous pollen must have evolved from gymnospermous prototypes, so that certain intermediate types must have existed; this is exactly what HUGHES (1961a) has demonstrated by his find of ovules associated with pollen of the *Eucommiidites* type, which suggest a chlamydospermous (*i.e.*, early angiospermous!) affinity. As HUGHES himself has emphasised, there is steadily mounting evidence (*e.g.*, the frequent occurrence of ephedroid sporomorphs) that in the Mesozoic chlamydospermous forms were wide-spread. In these angiospermoid Cycadopsids the other criteria of angiospermy may have developed sooner or later than the pollen character. Even recent taxa generally recognised as Angiosperms, such as certain Piperales, have 'non-angiospermous', or at least not typically angiospermous, pollen, and when this retained ancient feature is not the only primitive character but occurs in conjunction with such features as a vessel-less secondary xylem and a delayed development of the embryo (*Sarcandra*), the obvious conclusion is that such plants have not 'completed' the evolutionary sequence transforming a cycadopsid 'Gymnosperm' into an 'Angiosperm' and are, in fact, still hemi-angiospermous. This reflects the phylogenetic history of the Angiosperms, which must have started as early as or even before the Jurassic by the almost imperceptible initiation of 'Angiosperm trends' and continues to this day, but one must agree with SCOTT et al. and with HUGHES that, judging by the evidence from palaeobotanic records, in the majority of the evolutionary lineages leading to what we are wont to call 'The Flowering Plants' a more or less complete state of 'angiospermy' was presumably not attained long before the Cretaceous. One must, however, not disparage AXELROD's contention that Angiosperms may have originated in a tropical region and, after having reached the status of full-fledged Flowering Plants, migrated polewards to reach the more moderate climates by the end of the early Cretaceous. If the present distribution of the more primitive groups of Angiosperms may be taken to be relevant to this problem—and I believe it is—their occurrence in the tropics with a concentration or 'main massing' of the most primitive representatives in the East Asian–Western Pacific–Australian area must be significant and indicate that they have probably always been there, in other words, that they originated in the Old World tropics. Examples are Pandanaceae, Cyperaceae-Mapanieae, Dilleniaceae, Actinidiaceae, Magnoliales, *Eupomatia*, Monimiaceae, Chloranthaceae, Casuarinaceae. It is not at all improbable that some of these groups attained a high level of angiospermy before the Cretaceous, but fossil evidence from this area is much needed.

In any event, the evolution of the Angiosperms extended over a long stretch and it is impossible to draw a sharp dividing line between gymnospermous Protangiosperms and full-fledged Angiosperms—in some recent taxa the ultimate level of angiospermy is not even developed—so that the whole controversy around the 'age' of the Angiosperms is rather inane. One can state without too much exaggeration that the Flowering Plants are as old as one wants them to be, depending, among other things, on the choice of the criteria of angiospermy. The viewpoints regarding the ancestry and the relationships of the Angiosperms, if properly exploited, may help build up the morphology of the Flowering Plants from the features of the cycadopsid Gymnosperms.

13

The Outstanding Controversy—
Floral Theories

Historical Review. The Classical Theory. Critical Appraisal of
the Classical Concept. The Axial Theory. Requisites for a Comprehensive Floral Theory.

If one disregards C. F. WOLFF's (1769) discussion of the serial
homology of floral appendages with vegetative leaves, the first comprehensive morphological floral theory goes back to the German author, poet,
philosopher and naturalist J. W. GOETHE (1790). The views expressed
by GOETHE and some of his contemporaries became known as the 'idealistic' morphology, the leading principle being the postulated homology
of certain entities (organs, etc.) through their relation to a common basic
pattern, a theoretical prototype or 'idea'. GOETHE's floral theory was
based on several postulates (some of which were implicitly taken for
granted), viz., (1) all reproductive regions of the Angiosperms, the
'flowers', are homologous through being modified shoots, (2) all lateral
organs of such a modified shoot are consequently homologous with leaves,
and (3) some of these leaf homologues (called 'metamorphoses' in
GOETHE's terminology) are fertile and constitute the male and female
genitalia. GOETHE was also the first to point out the transitions between
perianth lobes and genitalia, mainly in proliferated flowers, and the
teratological conditions known as phyllody and virescence (antholysis),
as corroborative evidence of the foliar nature of the floral appendages.

In the idealistic morphology and its successor, the typological phytomorphology, an elaborate theory of the 'flower' was gradually built, on
these fundamentals, which was also suitable for descriptive purposes
(phytography, DE CANDOLLE 1826). The original theory has survived to
this day, and although it gradually became somewhat modernised the
basic assumptions have not been altered appreciably, even if an attempt
has been made to give it a phylogenetic interpretation. It is, in fact, the
only theory presented in the majority of the leading handbooks and
educational texts, which has given many a student the idea that it is indeed

the only morphological interpretation of the 'flower'. The appearance of the most recent treatment of the subject, EAMES' *Morphology of the Angiosperms* (1961), must be seen as the last attempt to stem the tide of modern phytomorphological thinking and as the closing of a period.

Having first been proposed for angiospermous 'flowers' and later extended to include the reproductive regions of pteridophytic and gymnospermous groups, it is a typical example of an 'Angiosperm-centred' morphological theory. It is variously referred to as the 'classical', 'typological', 'foliar' (appendicular), Ranalian or euanthium floral theory, or as the hypothesis of 'phyllospory' (LAM 1948, *et seq.*, and others, such as FAGERLIND 1958 and PANKOW 1962).

Although an 'axial' origin of the ovules had already been suggested earlier, one can state that the first substantial contribution based on this postulate was PAYER's study of the development of the placentae and the ovules (1859). His conclusions were disputed by VAN TIEGHEM (1875) who, on the ground of anatomical studies, claimed to be able to prove the foliar nature of the gynoecia. This controversy was not considered to be quite settled until about 1900, after ČELAKOWSKY in a series of papers had persuaded the botanical world to decide in favour of 'leaf-borne' ovules (*Blattbürtigkeit der Eichen*). Thus the classical dicta became once again universally accepted, so much so that the essential working hypothesis, the foliar interpretation of all floral appendages, came to be regarded as factual. LAM (1948), and MELVILLE (1962) summed up various discrepancies and inconsistencies of the traditional interpretation of the 'flower'. LAM repeatedly pointed out that the current views cannot possibly apply to all forms of angiospermous genitalia, so that he again defended the occurrence of axis-borne ovules in at least some angiospermous taxa. He was not the first to propose a polyaxial angiospermous flower, but his predecessors—chiefly adherents of the pseudanthium theory of the flower such as WETTSTEIN and HAGERUP—became entangled in a maze of ingrained traditional connotations and semantics leading them astray, or (like NEUMAYER) they created other stumbling blocks themselves. The pseudanthium idea was mainly employed in the field of phylogenetic botany (classification or systematics) and did not have much impact on purely morphological inquiries.

The third and last group of floral theories is based on an origin *sui generis* of the floral appendages or at least of the fertile ones. After what has been said in Chapter 4 regarding organs *sui generis*, it will be clear that I reject these theories whilst admitting that the sporangia and their semophyletic derivatives (nucelli, thecae and some sterile homologues such as integuments) are elements without truly sterile (vegetative) counterparts. Protagonists of the *sui generis* interpretation (GRÉGOIRE 1935, 1938, McLEAN THOMPSON 1934, 1937, PLANTEFOL 1948,

BUVAT 1952, 1955, LANCE and others, mainly representatives of the French histogenetic school) claim that a shoot apex in its generative phase has a structure different from that of the vegetative apex, so that the appendages formed at the floral apex are not comparable (*i.e.*, not homologous) with the lateral organs developing at the vegetative growing point. Indeed, several students of morphogenesis agree that the apex in its generative stage is more or less different from the vegetative phase, but others deny that there is a *fundamental* difference between the two phases. (For a summary, see HAGEMANN 1963.) The protagonists of the French school claim that there is a '*méristème d'attente*' in the vegetative apex which only becomes active in the generative phase, but most authorities on morphogenesis at the shoot apex and related subjects (*e.g.*, WARDLAW, GIFFORD and also some German workers such as ROTH) agree that morphological differences between the vegetative and the generative phase, if any, do not preclude the homology of the lateral derivations of vegetative and floral apices. There is also a fairly general consensus of opinion (the French morphogenetic school again excepted) that one can explain the formation of a floral apex as a conversion of a vegetative apex by the action of some florigenic stimulus which transforms the whole apex (and does not bring only the *méristème d'attente* to life, as the French school has it). The initiation of the stimulus may have different (alternative) causes, such as a change in photoperiod or an internal rhythm, often coupled with a phytohormonal induction or other physicochemical processes, effects of temperature and other external factors, but this does not concern the morphologists beyond the recognition that certain changes take place, resulting in a different pathway of the histogenesis (organogenesis in its proper sense) at the shoot apex.

Although the adult organs formed at the floral apex are more or less different from those developed on a vegetative apex, there is no cogent reason to reject the basic homology of the elements constituting the floral region with those of a vegetative shoot. However, it is necessary to find an appropriate formulation of this homology in general terms whilst avoiding overambitious claims. Adherents to the *sui generis* theories excepted, nobody will probably object to a morphological interpretation of the floral region as a 'shoot' consisting of a main axis supporting lateral organs (leaf homologues) *which potentially bear buds, i.e., secondary axial derivatives, in their axils*, as the basis of typological and phylogenetic inquiries into the structure, origin and semophyletic relations of the reproductive zones of the Cycadopsida. This is why I think that there is very little point in discussing CROIZAT's (1960) explanation of the origin of the angiospermous 'flower' in which no such basic pattern is recognised and certain standard principles of phytomorphology (*e.g.*, the homology concept) are violated. CROIZAT derives a 'flower' from ovules and 'scales', the latter being supposed to become 'sexualised' into stamens, etc., so that his 'theory' is nothing but a *sui generis* hypothesis in disguise and to my view altogether unacceptable.

After this digression in the histogenetic field, the two remaining alternative interpretations of the floral region can be appraised in terms of the same basic pattern, a modified shoot. In view of the *sui generis* nature of the sporangia and sporangial homologues, the difference between two groups of theories is best expressed by the distinction between *leaf-borne* sporangia (or leaf-borne sporangial homologues) and *axis-borne* sporangia (sporangial homologues), or (LAM's) 'phyllospory' and 'stachyospory', respectively. In spite of the difficulty of defining 'leaf' and 'stem', this distinction is generally considered to be unequivocal because the phylogenetic inquiry enables the retracing of fertile organs to their origin: if they are indeed derived from sporangium-bearing elements (telomes or syntelomes) that clearly had a semophyletic connection with foliar (lateral) organs, phyllospory can be accepted, but if the genitalia apparently had an independent semophylesis that roots in sporangium-bearing telomes or syntelomes which never developed into laminose assimilatory organs, their qualification as 'stachyosporous' seems appropriate. At the evolutionary level of the Angiosperms, trophophylls and 'stems' are not only clearly distinguishable by their morphology and by the leaf-axillary bud relation, but also by the mode of development of the leaves from lateral primordia of a shoot apex, and of the stems from the shoot apex itself or secondarily from axillary buds (which are potential shoots, *i.e.* axes). The organogenetic definition also holds true in those cases in which axillary buds are lacking as in Monocots: if additional stems are formed at all they arise from a di- or occasionally trichotomy of the shoot apex (Pandanaceae, Palmae, *Dracaena*, etc.), whereas the leaves develop from lateral primordia. The fundamental homology between vegetative and generative shoot apices thus becomes very significant. This is by no means a novel point of view, because it has been the basis of many morphological and histogenetic deductions, but a simple fundamental point has been overlooked or summarily dismissed, as I shall demonstrate presently.

Turning to the principal arguments adduced by protagonists of the classical theory, they are mainly derived from three sources, *viz.*, anatomical features, histogenetic (organogenetic) development, usually loosely referred to as 'ontogeny', and teratological malformations. A fourth, the so-called serial homology of trophophylls, transitional phyllomes (prophylls, etc.), sterile floral appendages and genitalia, cannot be accepted as having any demonstrative force, because the serial homology presupposes a development of a sequence of equivalent elements and this requires identical or near-constant morphogenetic conditions in the vegetative shoot apex and in the apex in its generative phase, which is still a debated issue, as we have seen. Even if one admits a serial homology of some degree, this must be interpreted in terms of the general pattern of the floral region, *i.e.*, of a shoot bearing lateral leaf homologues *subtending axial organs* (derivatives of axillary buds) and this is the point that is so often overlooked.

Anatomical data are often much abused in that a more or less successful *interpretation* of certain topographic constellations on the basis of a postulated foliar nature of the genitalia does not 'prove' anything if an alternative explanation based on a different assumption (MAJUMDAR 1956, MEEUSE 1964a) can be given. Some examples have been studied by MOELIONO, who has also found that figures and descriptions of the anatomical structure of the floral region are often incomplete or inaccurate and sometimes utterly misleading. Another relevant case is the interpretation of the leaf gaps in the stelar system of the floral region. The presence of such gaps above the point of insertion of stamens and carpels has been claimed to be a very cogent argument supporting the appendicular nature of the genitalia (TEPFER 1953, EAMES 1961), but it is only indicative of the presence of a 'unit' consisting of a lateral organ and its axillary associated organ (a bud or an axis) at the proximal side of each leaf gap. One often encounters dual units of an epipetalous stamen and the 'opposite' perianth lobe which develop from a common primordium (BENNEK 1958, MELVILLE 1962, 1963, SATTLER 1962). Both histogenetically and anatomically, the stamen and the associated perianth lobe agree with a fertile axis and its subtending bract, respectively. However, if only one element is present below a leaf gap there is no reason to assume that it is always the *axillary* partner that is lacking, so that the remaining organ is the leaf homologue. BREMEKAMP (1956) is one of the very few phytomorphologists who emphasise the (apparently) consistent lack of axillary buds in the androecia and gynoecia of the Angiosperms. This absence of associated axial organs is for him a sufficient reason to express considerable doubt about the homology of the genitalia with trophophylls, and I wholeheartedly agree that one should not make light of this point, if only for typological reasons. In terms of organogenesis at the shoot apex, the occurrence of a single organ associated with a leaf gap can be explained by assuming that it represents (1) the foliar organ, the axial partner having become reduced, (2) the axial organ, the subtending foliar bract having become vestigial, or (3) a complex element originated from a phylogenetic fusion (adnation) or amalgamation of the axis and its bract (MELVILLE 1960, 1962, 1963, MEEUSE 1963c; see also Chapter 14).

In spite of all previous claims, the anatomical arguments are anything but conclusive and statements to the effect that the anatomy of the floral region is in perfect harmony with an appendicular character of the genitalia are decidedly unwarranted.

The histogenetic development of the genitalia is often said to be so similar to those of the vegetative phyllomes that this is sufficient proof of the foliar character of stamens and carpels. If this be true, why is it that students of the organogenesis at the floral apex disagree? If the development of the genitalia at the floral apex were a clear-cut demonstration of their foliar nature, there would be no reason for disagreement or criticism, but a comparison of some recent publications on the same subject, the histogenesis of the gynoecia of the Primulales (ROTH 1959, PANKOW 1959, 1962, SATTLER 1962), shows that the conclusions are sometimes diametrically opposed. Various workers have found that, in the young androecial and gynoecial region of the floral apex, primordia develop which later divide, usually by the formation of an adaxial protuberance, and give rise to a 'unit' comparable to a cauline, adaxial organ and an abaxial subtending bract. The fertile elements of such units (stamens and placentae) are consistently derivatives of the *adaxial* primordium and consequently correspond with an axillary organ, which is of course of a cauline nature. Therefore, if the histogenetic differentiation at the generative apex is to be used as evidence at all, there are many instances in which it points indubitably to *axis-borne* anthers and ovules (MOELIONO 1959, PANKOW). This leaves the teratological arguments. The relevant malformations, cases of virescence and phyllody, belong to the category of irregular growth phenomena and not to the truly atavistic abnormalities (HESLOP-HARRISON 1952; see also p. 27), so that, whatever shape the teratologically developed floral appendages may assume, their morphology does not necessarily correspond with ancestral (primitive) stages in the semophylesis of the genitalia. Antholysis of the floral region has frequently been observed in plants infected by a 'leaf-virus' causing lesions and other symptoms *in the leaves* (BOS 1957). This virus-induced malformation of the floral region can only be caused by an inadequate florigenic stimulus which is not capable of a complete transformation of the shoot apex into its generative phase, so that the vegetative development of the common primordia of a unit of a lateral organ and its associated axillary organ is stronger than in the normal floral apex of a healthy plant. The more 'vegetative' development (virescence and phyllody) of the floral apex becoming relatively stronger as the leaves are more diseased, the weakening of the florigenic stimulus is compatible with the supposition that this stimulus is essentially a translocation of certain morphogenetic substances, such as phytohormones, towards the apex (MEEUSE 1963c). Other antholytic 'degenerations' of flowers are most probably also caused by irregularities in the normal processes contributing to the florigenic stimulus and, accordingly, do not prove the homology of the genitalia with leaves.

The classical floral theory thus appears to be built on very weak foundations, so that the lack of supporting phylogenetic evidence—the traditional annoying absence of fossil Angiosperm ancestors—is most significant, even if not unexpected. As I shall repeat *ad nauseam,* the traditional 'sporophyll' concept, the preconceived homology of the genitalia of the Angiosperms with phyllomes, lateral (appendicular) organs or in plain words 'leaves', later extended to include the reproductive organs of other cormophytic groups, was so fundamental in plotting a fixed course in morphological thinking that taxonomic and phylogenetic inquiries were similarly biased. This 'established' certain notions which are untenable and led, among other things, to such absurdities as the wild goose chase for Angiosperm ancestors with the postulated sporophylls (see MEEUSE 1965, pp. 158–75).

As I shall attempt to demonstrate, the alternative postulate of axis-borne cupulate ovules opens up many new approaches to morphological, taxonomic and phylogenetic problems, some of long standing, and renders highly probable various relations previously deemed 'impossible'. In this chapter, only the general aspects will be discussed.

The ideas expressed by PAYER were based on rather crude observations and need not be discussed in detail here. Apart from the pseudan-thium theory, the phytomorphological consequences of which were not always understood by its own protagonists, no comprehensive theory based on an axial or partly axial origin of the genitalia was published before 1924, when NEUMAYER published his emended version of the pseudanthium theory (discussed in more detail in Chapter 10). He postulated a common biaxial morphological pattern of the floral region, the lateral axes or gonoclads being supported by bracts, and assumed that each gonoclad bears a sporophyll which is (at least in the Angiosperms) so vestigial as to be almost non-existent. For practical purposes, his suggestions amount to a derivation of the genitalia from an axis subtended by a bract, so that a carpel must be interpreted as a combination of this bract (a sterile phyllome) with its axillary fertile organ. Similar ideas were expressed by HAGERUP, who called the sterile appendicular components of the gynoecia 'pseudo-carpels'. HAGERUP, like NEUMAYER, felt obliged to locate a sporophyll somewhere, because the ingrained classical dicta require leaf-borne ovules, but he presumably did not sense a corruption of conventional phytomorphological principles in his own teachings when he indicated the integument as the sporophyll. LAM unequivocally defined 'stachyosporous' Higher Cormophyta as plants with axis-borne sporangia (or sporangial homologues) and lacking true sporophylls ('eucarpels'). If axillant bracts of ovuliferous axes ('stegophylls') participate in the formation of a gynoecium they are pseudo-carpels, which are fundamentally different from what he calls eucarpels (eusporophylls), supposed to correspond with the classical carpel (megasporophyll). Although I cannot support the occurrence of 'eusporophylls' in the Higher Cormophyta, it is clear that LAM's proposals are conceptually quite in order. How easy it is to confuse principles and semantics is shown by PANKOW's paper of 1962 on the histogenesis of gynoecia. Although he concurs with LAM in accepting both phyllospory and stachyospory in the angiospermous ovaries, he expressly states in a footnote that he does not distinguish carpels from pseudo-carpels, which shows a lack of appreciation of the relevant fundamentals.

In 1960, MELVILLE proposed what he claims to be 'a new theory of the Angiosperm flower'. In fact, his working hypothesis boils down to a modified version of NEUMAYER's floral theory and LAM's concept of stachyospory. MELVILLE postulates that all angiospermous genitalia are derivatives of 'gonophylls', *i.e.*, of units consisting of a bract-like organ subtending a fertile axis. The use of the term 'gonophyll' is deplorable, not only because it is semantically ambiguous, but also because it had been used by NEUMAYER in a different sense. One of the corner-stones of MELVILLE's theory is the presumption that the fertile axial element of a 'female gonophyll' is bifurcate. This enables him to give a very convenient interpretation of the peculiar mode of vascularisation of the carpels of the Polycarpicae, it is true, but the occurrence of such bifurcate fertile axes is based on precariously slender phylogenetic evidence. MELVILLE's typological comparison of certain types of genitalia is unwarranted in such cases as the gynoecia of *Gnetum* and the carpels of *Magnolia*. Some anatomical features in the fertile zone of the floral axis are badly represented (*Magnolia*) or too stereotype (*Caltha*). Another objection to the 'gonophyll theory' is the fact that gonophylls are directly derived from 'primitive gonophylls' which are sporangium-bearing, the origin of integuments, arilli and other accessory organs of the sporangia remaining unexplained. This last argument is equally cogent in querying the phylogenetic derivation of an angiospermous carpel from a primitive sporophyll which, according to ZIMMERMANN's (and LAM's) method of representation, is a mixed syntelome and bears sporangia. Such suggestions of a phylogenetic relation can be accepted only if they are augmented by a plausible morphological interpretation of the integuments, arilli and other organs associated with the ovules. Angiosperms have essentially *cupule-borne* ovules (MEEUSE 1964a), and all floral theories must be compatible with this fundamental structural pattern. The cupule originated from the fusion of phyllodic protocauline telomes or syntelomes independently of the trophophylls and is not a 'leaf', hence the cycadopsid ovule is not leaf-borne.

The outcome of this brief review is that, although the traditional tenets appear to be untenable, thus far no adequate comprehensive 'axial' theory has been proposed in substitution for the classical hypothesis. Obvious non-conceptual requisites of the alternative working hypothesis are: (1) an interpretation of the peculiar laminose stamens and the carpels of the Polycarpicae and some other groups; (2) an explanation of the origin of the organs associated with the ovule, such as the funicle, the integuments and the arillus; (3) a common standard pattern of the floral region of the Angiosperms to be used as the basis of a revised typology of the angiospermous reproductive zones, mainly for descriptive purposes in practical systematics; (4) a phylogenetic approach to the morphology of the fertile region of the Angiosperms, thus linking it semophyletically with the reproductive structures of other spermatophytic groups; and (5) the co-ordination of (3) and (4) so as to obtain uniformity in terminology and interpretation.

14
Pseudo-Phyllospory

Organs of Intermediate Nature. The Fundamental Equivalence of
Vegetative and Floral Apices. The Proterogenic Merging of Unit
of Fertile Axis and Bract. Phylogenetic Consequences of the Theory
of Pseudo-Phyllospory.

The competing postulates concerning the morphological nature
of the genitalia of the Angiosperms have been discussed in Chapter 13,
and we have seen that the 'stamens' and 'carpels' are either defined as
leaf-homologues or as organs wholly or partly of axial derivation, if
they are not regarded to be organs *sui generis*. Only the working
hypothesis based on the premise of axis-borne (properly speaking,
cupule-borne and coaxial) sporangia will be considered here and the
origin of the carpellate angiospermous ovary is explained as a combination of fertile axes (gynoclads) with sterile supporting bracts (stegophylls
or 'pseudo-carpels'). Such combinations have been interpreted by MEL-
VILLE (1960) as adnations or concrescences of the fertile and the sterile
elements, which implies that the constituting components to a large
extent retain their individuality and characteristics.

LAM, who originally (1948–1954) divided the Flowering Plants into
'phyllosporous' and 'stachyosporous' forms and stressed the fundamental
difference between these conditions throughout the Cormophyta, later
(*e.g.*, 1961b) postulated the existence of an 'intermediate' group of
Angiosperms supposed to have genitalia which are intermediate between
the phyllosporous and the stachyosporous conditions. MAJUMDAR (1956)
had previously also suggested an intermediate character of angiospermous
genitalia, but did not give it a phylogenetic interpretation, as LAM attempts to do. LAM explains, rather vaguely, that the older Cormophyta
had phyllosporous and stachyosporous tendencies and that all through
the evolution of the Flowering Plants such intermediates with 'mixed'
tendencies persisted. This recent emendation of his theory is not compatible with an independent origin of reproductive organs with phyllome-
borne and axis-borne sporangia; in fact, it contradicts his own teachings
of the fundamental—phylogenetically speaking, *ab initio* established—dif-

ference between 'phyllospory' and 'stachyospory'. I suspect that the difficulty of finding satisfactory criteria of phyllo- and stachyospory had something to do with this change of view, which is connected with the admission (LAM 1959) that the difference between leaf and stem ultimately escapes us. By resigning himself to this somewhat defeatist attitude, LAM weakened his own argument, that 'phyllospory' and 'stachyospory' are fundamentally different alternatives, to an inconclusive one.

The palaeobotanic evidence appears to leave room for only one working hypothesis, viz., the derivation of all gymnospermous and angiospermous plants from the Progymnospermopsida; and since these ancient Cormophytes had axis-borne sporangia, all derived groups are essentially 'stachyosporous'. That the fertile organs in several (mainly cycadopsid) groups such as Cycas and Magnoliales are more or less 'leaf-like' in appearance is beyond doubt, so that the contention of MAJUMDAR and LAM that organs of an 'intermediate' nature occur is not altogether without ground. LAM's phylogenetic interpretation obviously being unacceptable, another explanation of the intermediate condition must be sought. The semophyleses of the ovules, gynoecia and male synangia (stamens) will be discussed in other chapters, so that only one aspect need be treated here, viz., the association of organs belonging to different morphological categories into close-knit complex units.

The study of flower induction has taught us that one can explain the initiation of the morphological changes of a floral apex as the result of a 'florigenic stimulus' moving into the apex. As long as the manifestly changed morphogenetic forces only result in a different growth regulation, the primordium at the receiving end, although yielding a phenetically different organ (i.e., an organ with an altered 'Gestalt'), does not develop into an organ of a fundamentally different morphological category. An appendage formed at a floral apex may be expected to belong to the same principal category as the organ that would have developed from the same primordium if the florigenic stimulus had not changed the regulation of the morphogenetic processes. Among the Angiosperms there are several examples of photoperiodic taxa, i.e., of forms in which flowering initiates only if the length of the alternating periods of illumination and darkness fulfils certain requirements. Under appropriate experimental conditions, such plants can be made to flower at will; and this means that a florigenic stimulus acts upon a vegetative apex that would have produced a vegetative shoot if the length of the photoperiods had not been readjusted, and transforms it into a floral apex. I think that this reasoning may be extended to include other possible forms of induction of flowering and one may assume that, generally speaking, a potentially vegetative shoot apex is altered by the action of some florigenic stimulus, irrespective of the mode of induction of the florigenic stimulus (day-length, effects of temperature changes, or an internal regulation mechanism). This agrees with the fairly generally held opinion that the floral apex and a vegetative shoot apex as well as their mature derivatives, a 'flower' and a leafy branch, are fundamentally equivalent. The morphological equivalence can be extended to the primordia of the apices and the appendages of the mature axes, respectively. This is the essence of an interpretation of the floral region which, I believe, is accepted by the majority of contemporary phytomorphologists.

From the moment a shoot apex enters the phase of a reproductive apex (and becomes a 'transition apex', in WARDLAW's terminology), several visible and invisible changes take place, ranging from a different mode or rate of cell division resulting in an altering shape of the floral apex to a measurable difference in biosynthetic processes (BONNER and ZEE-VAART 1962, GIFFORD and TEPPER 1962), but these changing conditions and the subsequent differentiation of the reproductive tissues are not regarded as of the most fundamental. However, the most striking difference in the organogenesis of the floral appendages is that often a fertile organ (a 'carpel' or more rarely a 'stamen') originates as a single primordium which does not become divided into two associated primordia, whereas in the vegetative apex a leaf primordium usually forms an adaxial derivative, the origin of the axillary bud. There are numerous examples of stamens and gynoecia which originate from single primordia that split into two, of which the adaxial one becomes a stamen or a placental area, and the abaxial one a perianth lobe (often a petal) or a segment of the ovary wall, respectively. The equivalence of the floral apex and a vegetative growing point includes an equivalence of their primordia, which means that a 'unit' of two primordia derived from a common initially single primordium represents the primordia of a phyllome (or bract) and its axillary associated axis, in other words, epipetalous stamens and placentae are clearly axes supported by bracts. The development of an appendage of the floral axis from a *single* primordium that does not divide into two primordia must be interpreted as a precocious simultaneous development of a potential precursor of a fertile axis and its subtending bract, the splitting of the primordium (the precursor) having become suppressed. The undivided primordium and its derivative clearly represent a dual entity, a 'unit' of a lateral organ (bract) and an axis.

The clear-cut cases of a separate development of epiphyllous stamens and the opposite perianth lobe, or of an ovuliferous placenta and the opposed segment of the ovary wall, are considered, by such authors as LAM (1948, etc.) and PANKOW (1962), to be indications of the 'stachyo-spory' of the taxa exhibiting this type of histogenetic development. PANKOW even believes that this form of histogenesis provides the criterion of stachyospory and that the alternative, the development of a carpel from a single primordium, is the organogenetic manifestation of phyllospory, but he cannot prove his own point because he himself found doubtful (transitional!) cases (*Myosurus, Alisma*). In fact, there are many intermediate stages and, even in the extreme case of the apocarpous gynoecia of the Polycarpicae, the single carpel primordium has a differentiation on the adaxial side, the '*Querzone*' of the school of TROLL.

The obvious corollary is that in *all* angiospermous taxa there is a fundamental unit of a fertile axis and its axillant phyllome (a gonoclad subtended by a bract or stegophyll) which is frequently clearly recognisable as a dual unit both histogenetically and in the fully developed stage, *e.g.*, a petal and an epipetalous stamen, derived from the same primordium (in Rhamnaceae: BENNEK 1958; in Caryophyllales, Primulales: SATTLER 1962), but sometimes initiates as a single element that does not divide before differentiating into a mature appendage (*e.g.*, the carpels of many and the 'stamens' of some Polycarpicae). In the latter case the full-grown appendage is an organ of a truly 'mixed' nature, combining characteristics of a fertile axis and its bract in a peculiar fashion which is not an adnation as MELVILLE has it, but more in the nature of an amalgamation. The diverse modes of development of a bilaterally symmetric, dorsiventral phyllome and a terete (radially symmetric) gonoclad are undoubtedly more or less antagonistic, so that, if the morphogenetic forces inducing a phyllomic and an axial differentiation act simultaneously on a single primordium, the resultant of the combined determinative stimuli may be expected to produce such singular growth processes that the full-grown organ, though exhibiting characters of both lateral and axial organs, acquires some novel features. Examples of such singular characteristics are provided by the laminose 'stamens' of several Magnoliales and *Nymphaea* and the carpels of apocarpous Polycarpicae and Leguminosae. The anthers are, to my mind, derived from stalked synangia and consist originally of (usually four) intimately fused microsporangia, but in Magnoliales (and in *Nymphaea*) the thecae are often segregated (they may even be marginal!). Contrary to the current interpretation, my own view is that this is a *secondary* separation of the thecae, *i.e.*, a splitting of the synangia, caused by the interaction of opposing determinative forces, the tendency towards the development of a bilaterally symmetric organ (the phyllome) apparently predominating over the tendency to form a terminal synangium. The vascularisation of the androclad sometimes also became split, so that a typical laminose stamen has three traces, the median one corresponding with the main trace of the incorporated stegophyll and the two lateral ones with a morphogenetically split androclad (= staminal) trace (see Figure 8). Similarly, the carpels often have three traces, one (the dorsal) representing the main trace of the bract-component and the two laterals the divided gynoclad (placental) trace. Significantly, the leaf-gaps associated with such stamens and carpels are usually more or less normal in comparison with a gap in the stelar anatomy of the vegetative region, but the adaxial bundle of the leaf-trace of a fertile appendage frequently bifurcates before leaving the floral axis and enters as the paired lateral vascular supply to the fertile elements of the stamen or carpel.

The postulated morphogenetic merging of a gonoclad-bract unit I have called 'pseudo-phyllospory' for historical reasons, but in the gynoecial region this is only the extreme case of a long series of intermediate stages: there are various 'degrees' of pseudo-phyllospory. Pseudo-phyllospory of stamens is apparently rare and seems to be confined to representatives

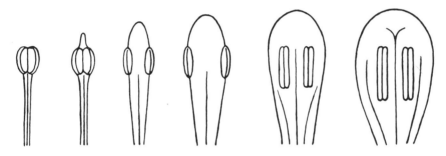

FIGURE 8. Semophylesis of the pseudo-phyllosporous stamen. The development is from a stamen with discrete anther to the laminose type of the Magnoliales and Nymphaeales (bract of stamen omitted in prototype).

of the Magnoliales and a few other Polycarpicae (including *Nymphaea*). The phylogenetic interpretation of its origin is not so easy to explain because it is not a simple adnation or concrescence of organs, but a much more intimate union. I have suggested (MEEUSE 1963c) as a possible cause a semophyletic contraction of a fertile region into a flower, which resulted in a much closer proximity of the appendages and also in different interactions of morphogenetic processes. One could also call it a neotenic or proterogenetic development of the undivided floral primordium, resulting in the precocious development of the axillary fertile organ before it became segregated by the splitting of the common primordium; but such terms are rather meaningless without a suggestion of a possible causal mechanism.

Transitional phases are rare (*Salix?*), but they suggest that the axillary gynoclad tended to become adnate to its stegophyll and that the adnation gradually shifted to younger developmental stages as the result of morphological changes in the floral region, presumably concomitant with and influenced by the positive selective effect of a better protection (and feeding?) of the ovules.

The working hypothesis of pseudo-phyllospory is compatible with several phenomena which have traditionally been the corner-stones of the appendicular (foliar) floral theory, particularly with occurrence of malformations. In the floral regions of plants with virus-produced lesions in the leaves, antholytic changes often occur which become more pronounced as the virus attack becomes more severe or, what amounts to the same thing, as the plant is infected in a younger stage of development (Bos 1957). The virescence and phyllody induced by a leaf-virus can only be explained by an alteration in the biochemical processes in the infested leaves leading to a diminution of the florigenic stimulus (one could think of the competitive synthesis of new virus-particles which withdraw building stones of nucleic acids or their precursors from the normal biosynthetic processes, so that such substances—if associated with a flower-inducing stimulus—are not available for other purposes). The falling-off of a florigenic stimulus inevitably results in a suboptimal induction of flowering and, in view of the transformation of a vegetative apex into a floral apex, this means that the apex more or less 'reverts' to the vegetative phase and the primordia of the fertile appendages develop in a more 'vegetative' direction, *i.e.*, they become more leaf-like in appearance and may even turn green. In this connection, it is highly significant that in extreme cases of virus-induced phyllody of the gynoecium the carpels develop as leaf-like elements *with an axillary organ* which bears reduced (degenerated) ovules (*e.g.*, in *Tropaeolum*: Bos). If such teratological cases have any demonstrative force at all, the antholysis of a carpel into a green, leaf-like sterile appendage and an axillary ovuliferous organ indicates that the ovules are axis-borne rather than leaf-borne and that the carpel is an organ of mixed nature compounded of a merged bract-gonoclad unit.

Other ('spontaneous') teratological degenerations of the floral region can be explained, in a similar way, as irregularities in the florigenic stimulus; their interpretation must be based on the same principles. This reasoning can be reversed by assuming that perhaps the development of pseudo-phyllospory originated as a somewhat abnormal case (a 'hopeful monster') that gradually became predominant, and that natural selection did most of the rest, but this is conjectural and difficult to prove. In some cultivated 'sports' of Japanese *Prunus* species, the gynoecia consist of a whorl of very leaf-like sterile organs and, because all (vegetatively propagated) specimens of such cultivars exhibit this gynoecial phyllody in all their flowers, it seems to be a genotypically controlled character. The florigenic stimulus is apparently inadequate to form fertile carpels and such cases when arising in nature are of course doomed to disappear for lack of progeny, but they are illustrative in that they demonstrate that changes in the biochemical control mechanism of floral induction may have profound effects on the morphology of the reproductive region, conceivably also in an evolutionary sense.

Pseudo-phyllospory is a derived condition, and the occurrence of pseudo-sporophylls may be taken as an indication of a relatively advanced phase in the evolution of the reproductive organs. If the ovuliferous scales of the coniferous cones are disregarded (which does not mean that pseudo-phyllospory did not play a part in the semophylesis of these complex organs), pseudo-phyllospory is apparently restricted to the Angiosperms, with the exception of the female specimens of *Cycas*. Both the Magnoliales and the female *Cycas* exhibit advanced phases of pseudo-phyllospory, so that the laminose 'stamens' of the Magnoliales and the pseudo-megasporophyll of *Cycas* are more advanced than the more common type of angiospermous stamen and the female cycadalean strobili, respectively. This has far-reaching consequences, because the genitalia of the Magnoliales and the female *Cycas* are reputedly the most primitive of their kind, a tradition reflected in various typological 'derivations' and in current systems of classification. I have discussed the morphology and phylogeny of the Cycadales in detail elsewhere (MEEUSE 1963a) and concluded that *Cycas* is one of the most specialised forms of its alliance. The typological derivation of the biovulate appendages of the female cone-like gynoclads, the so-called 'megasporophylls', from the pseudo-megasporophyll of *Cycas* (which must be interpreted as a merging of a *gynoclad*, the homologue of a *whole* female cycadalean cone, with its bract), which is found in many text-books and manuals, is undoubtedly fallacious. Similarly, the derivation of the stamens of the Angiosperms from a laminose magnoliaceous prototype is erroneous. Not only is the laminose type of stamen with separated thecae an 'advanced' condition, but it is also an organ of dual nature and represents a combination of a microsporangiate axis and its bract, whereas the 'normal' type of stamen with a filiform filament and a compact bithecate anther is wholly axial. The typological derivation and the semophyletic sequence must be read in the *opposite* direction, as indicated in Fig. 8.

The corresponding systems of classification based on a magnoliaceous prototaxon must also be rejected if one aims at a truly phylogenetic system. It cannot be denied that the Magnoliales exhibit several primitive features, such as a vessel-less secondary xylem and monosulcate pollen grains, but their extreme pseudo-phyllospory eliminates them at once from the list of prospective progenitors of all other Angiosperms or even of all Polycarpicae. The Magnoliales are relict forms which are the surviving over-specialised terminal groups of an early offshoot of an evolutionary line leading to some or all other dicotyledonous taxa, in other words, they constitute an evolutionary cul-de-sac.

The occurrence of carpellate pistils in various not closely related major taxa of the Angiosperms is indicative of the polyphyletic development of this condition in the independent evolutionary lines leading to the principal groups constituting the Angiosperms. Carpellate ovaries are found in the Polycarpicae except *Nelumbo* and the Laurales, in the Saururaceae but not in the other Piperales, perhaps in the Salicaceae but not in the majority of the Monochlamydeae, in the Liliales, Scitamineae, Commelinales, Juncaceae and several derived groups (Orchidales, etc.) among the Monocots, in the genus *Freycinetia* among the Pandanales, the remaining taxa of the same groups generally possessing 'naked' gynoclads bearing chlamydote ovules. The ovaries are apparently consistently carpellate in the Centrospermae and in the large Rosiflorae-Guttiferae-Parietales plexus with all its dialypetalous associated and sympetalous derivative groups. The 'scattered' occurrence of carpellate ovaries indicates that this condition developed independently more than once, so that it is not an exclusive feature of one or more of the larger taxonomic groups.

LAM's contention (upheld by PANKOW) that there are stachyosporous and phyllosporous Angiosperms cannot be reconciled with such a distribution of the carpellate ('phyllosporous') ovaries among the Flowering Plants—especially the occurrence of the 'phyllosporous' Saururaceae and the genus *Freycinetia* among 'stachyosporous' taxa, and of the ecarpellate lauraceous pistil in a group which is mostly carpellate, are difficult to explain away. The alternative hypothesis, based on the assumption that all cycadopsid Spermatophytes have axis-borne ovuliferous cupules (arils) and that cupuliferous axes (placentae) sometimes became merged with their subtending bracts, seems to me to be more compatible with the accumulated palaeobotanic, morphological, histogenetic and teratological data than any other theory alone and opens up the way to a semophyletic derivation of carpellate Angiosperms from non-carpellate cycadopsid ancestors.

15
Phylogeny of the Megasporangium: I. The Ovule

Ovules arc found only among the Spermatophyta, the descendants of the Progymnosperms. Since HOFMEISTER's time, the ovule has been recognised as the derivative of a megasporangium with the restriction that, properly speaking, only the innermost massive portion of the ovule, known as the nucellus, is truly homologous with a sporangium, the organs surrounding the nucellus being accessory structures somehow 'acquired' by the sporangium and thus making it an ovule. The definition of an ovule accordingly becomes simply: *An ovule is a megasporangium or its homologue surrounded by at least one associated covering organ.* There can be as many as three individual enveloping covers in the most advanced cycadeoid taxa; apparently the megasporangium first became an ovule with a single protective organ or integument, later, in some or all genealogical lines, becoming covered with a second and eventually a third associated organ. The first point to be decided is if the various elements constituting an ovule belong to a limited number of categories of homologous organs, a question which is closely linked with the inquiry into their origin. This is a very fundamental issue in connection with the interpretation of the female genitalia of the higher cycadopsid groups including the Flowering Plants and with its bearing on related subjects such as the advent of angiospermy and the phylogeny of the Angiosperms.

To begin with the sporangial element, the nucellus: the question of its origin has already been touched upon in a different context (see, *e.g.*, p. 37) and an origin *sui generis* has been postulated. The homology of all megasporangia and their derivatives follows from the vital function in the life-cycle, each sporophytic generation being produced by the fertilised gametangia of the female gametophytes of the previous generation, the gametophyte by the megaspore, and the megaspore by the sporangium. All vegetative organs are expendable (at least, one at a time), but as there can be no gap in the continuous alternation of sporophytic and gametophytic generations the sporangia are essential to maintain the species.

The nature and origin of the covering layers has been the subject of many discussions and speculations (see, *e.g.*, ROTH 1957, LAM 1959, MAHESHWARI 1960, CAMP and HUBBARD 1963b, MEEUSE 1963b, 1964b for some recent summaries), but most of the relevant suggestions were based on erroneous interpretations of the ovules and their stalks (funicles) or on Angiosperm-centred semantics. Typical old-fashioned ideas are, for instance, the identification of an integument with an indusium (a suggestion which can be ignored because many examples of vascularised integuments are known), with a 'leaf' or phyllome (making the funicle a leafy shoot and the ovule something like a terminal bud—which is absurd), or with a 'sporophyll' or a portion of a 'sporophyll' (an idea based on the postulate that the ovules are leaf-borne—but the term 'sporophyll' also being an example of confused semantics, the identification of an organ associated with or in close proximity to an ovule with a 'sporophyll' is a more or less forced interpretation to save the theory!).

An origin *sui generis* or *de novo* of the ovular coats has also been proposed, sometimes, perhaps, for lack of a better explanation. This idea I consider to be unacceptable, because, as I shall attempt to demonstrate, a semophylesis can be reconstructed for each protective organ corresponding with one or several of the orthogenies of major spermatophytic taxa.

In the past, the main difficulty has been the want of a suitable prototype, so that only a neomorphological approach is likely to provide acceptable proposals concerning the origin of the integuments. The first tangible result was BENSON's synangial hypothesis of 1904, which, although based on good palaeobotanic evidence, was not universally adopted. However, it was never altogether abandoned (it is accepted, *e.g.*, in TAKHTAJAN 1959 and MAHESHWARI 1960). An alternative suggestion on the origin of integumentary coverings, *viz.*, a derivation from aggregates of sterile telomes, has been made by WALTON (1953), who is followed by, *e.g.*, ZIMMERMANN (1959) and LAM (1959). A similar explanation is given by CAMP and HUBBARD (1963b).

The phylogenetic inquiry can be fruitful only if the prototype is 'primitive' enough, and this requires an ancient form with *naked* sporangia that is also at the base of a phylogeny leading to groups with tegumented ovules. The postulated origin of all gymnospermous groups from the Progymnospermopsida provides the most likely and suitable starting point to build up a semophylesis of the ovule (MEEUSE 1963b).

Although in this chapter only the 'phylogeny of a single feature' (LAM 1935), the evolution of the ovule, is dealt with, one must bear in mind that sporangia and ovules were originally always produced in large numbers on complex organs, the semophyleses of which are discussed in Chapter 11. The archetype of these multisporangiate organs is a rather complicated structure which shall be referred to as a 'mixed frond' for lack of a better name; the term is used only descriptively and is not intended to suggest that the sporangia are leaf-borne (see Figs. 1, 5, and 6). The fertile ultimate segments of the mixed fronds are structures which developed directly from telomes and 'are older than the leaf', so that their independent origin and general morphology, in conjunction with the apparent lack of an assimilatory function, justify their classification among the protocaulomes, *i.e.*, among potentially axial organs. These ultimate ramifications of the complex syntelome (the 'mixed frond') bore the sporangia (or ovules, as we shall see), sometimes on individual sporangiophores (as in *Archaeopteris*); but, in other cases, these sporangiophores tended to fuse and thus the sporangia (or ovules) became aggregated into closely packed groups (Aneurophytales). I shall refer to such groups as *synangia* irrespective of the degree of fusion of the sporangia, because there are all possible transitions from a group ('sorus') of associated but free sporangia with a common stalk (the common sporangiophore or synangiophore) to sporangia coalesced at the base and, ultimately, to the intimately fused synangia constituting, for instance, the male reproductive organs of the Neuropteridales (Whittleseyinae). The common 'stalk' of the synangium is also a syntelome without any attribute of a phyllome and hence indubitably referable to the potentially axial organs. It is important to note that neither this synangiophore nor its supporting axis (rhachis) bears any functionally foliar (assimilatory) appendages. The comparative morphology of whole progymnospermous fertile fronds, of the fertile fronds of seed ferns and of the female reproductive organs of the Cordaitales provides the semophyletic connections between these fertile regions (MEEUSE 1963b; see also Chapter 11). The differences between the sporangium-bearing structures of the early Gymnosperms and their progymnospermous progenitors are chiefly the oligomerisation of the number of megasporangia and the appearance of integuments in the derived groups. In conjunction with the arguments of BENSON (1904) and DE HAAN (1920) that the (outer) integument shows indications of being compounded of a number of equivalent 'units' (in seed ferns often recognisable as the free apical lobes of the integument), and BENSON's ingenious interpretation of the hollow lobes of certain pteridospermous integuments as reduced ('sterilised') sporangia, the obvious corollary is the homology of a protected (tegumented) ovule with a gynosynangium, the stalk or funicle representing the synangiophore, the integument a group of sterile fused megasporangia adapted to other functions, and the nucellus the only remaining fertile megasporangium (see Fig. 10). However, the argument remains the same if, instead of megasporangia, primitive unitegmic ovules after fusion and reduction are supposed to surround a central fertile ovule. The question arises whether all integuments developed in the same manner. I believe that the 'synangial' origin can only be accepted for the *outer* protective ovular covering, the outer or second integument, of the Spermatophyta. The inner integument must be older and was conceivably already differentiated in the Progymnosperms, most probably by the invagination of the sporangiophore (see Fig. 9). The gymnospermous ovule had always consistently been supposed to possess only one integument (except in the Gnetales) until the late and lamented W. H. CAMP pointed out that this is not correct; and indeed, a recon-

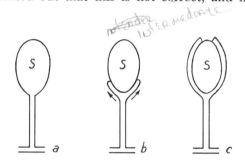

FIGURE 9. Origin of the inner integument as a circumvallation of the sporangium (S) by an outgrowth of the sporangiophore: *a*—'Naked' sporangium. *b*—Intermediate stage. *c*—Integument almost completely encasing the sporangium.

sideration (MEEUSE 1964b) reveals that the bitegmic condition is universal. The traditional integument is sometimes the outer one (in the Lyginopteridales), sometimes the inner one (Ginkgoales, Taxales, Pinales), and in other cases a fusion product of the two (Neuropteridales, Cycadales). The outer integument occurs among the Coniferophytina in several modifications, such as the fleshy outer layer of the ovule in *Ginkgo* and *Taxus* (in the latter called the 'aril'), the 'ligule' in *Auracaria*, the 'epimatium' of Podocarpaceae, the seed wings of the Pinales, etc.

As regards the functions of the ovular coatings, an accessory organ of this kind undoubtedly provided an additional protection of the vital megaspore and the vulnerable gametophyte against the deleterious effects of excessive drought, dampness, or solar radiation, and against damage by animals, the last factor becoming more important as pollination by animal agency became a major selective factor. However, this is not the only 'function' or, in a perhaps more accurate terminology, adaptative evolution of the integuments. In the first place, the tegumentary elements must have a physiological function in connection with the metabolism of the megaspore and the developing gametophyte, *e.g.*, as a source of nutrients, phytohormones and water. In several groups of the Spermatophytes the living contents of the integument are used up during the development of the gametophyte and the embryo, which points to another function, that of storage tissue for the gametophytic generation. Another adaptation is connected with the process of 'pollination', a term that I shall employ for the transfer of all cormophytic microsporomorphs, *i.e.*, microspores and, later, pollen, to the immediate vicinity of the megasporangium. In some Lyginopteridales (*Gnetopsis, Calatho-*

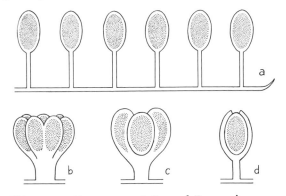

FIGURE 10. Diagrammatic representation of BENSON's synangial hypothesis of the origin of an integument: *a*—A number of coaxial megasporangia. *b*—The association of the sporangia into a sorus or synangium. *c*—All sporangia but the central one have become sterile. *d*—The sterile remains of the sporangia have completely fused to become an integument. *Note:* This derivation is presumably the explanation of the advent of the *outer* integument, which implies that the latter is not the derivative of an aggregate of sterile megasporangia but of reduced primitive unitegmic ovules.

spermum, etc.), the inner integument developed a distal and sometimes appendaged extension, obviously for the reception of the microspores. The distal extension and its appendages together constitute what is known as the salpinx.

In these forms, of Lower Carboniferous age, the outer integument was usually more or less deeply lobed so that it did not form a micropyle and apparently did not contribute materially to the catching of the male spores. In the more advanced seed ferns such as the 'Lagenostomales', the lobes of the outer integument were more connivent, and they formed a micropyle leading to the pollen chamber. The microspores were probably first caught by the lobes of the integument but were subsequently sucked in by a pollination drop exuded by the salpinx. The development of a pollen chamber is an important evolutionary trend because it initiated the advent of siphonogamy, which eventually led to double fertilisation and angiospermy, but the pollen-receiving function of the inner integument is still manifest in Corystospermaceae, Chlamydospermae and even Juglandales (MEEUSE 1964b).

The integuments sometimes develop into a fleshy (or dry, spongy to fibrous or corky) tissue, or form wings which are adaptations to various forms of dispersal (zoochory, anemochory, transport by water currents, etc.). Early examples of such adaptations are the winged integuments in Cordaitales and Pinales and the often massive ovular coats of neuropterid Pteridospermae (*Pachytesta*) and Cycadales. Some or all of the functions of the first protective layer can, *mutatis mutandis*, be performed by the second or third enveloping organ of the ovule.

The second protective organ of the ovule or outer integument is present in all gymnospermous groups and also in many Angiosperms. In the Angiosperms, the absence of a second integument in certain orders (*e.g.*, Juglandales) is only apparent, because whether or not one recognises an ovular coat as an integument depends on the interpretation of the gynoecia (MEEUSE 1964b); but in other groups (many Sympetalae), the integuments are often reduced and perhaps sometimes fused into one, or one of them may be rudimentary (rarely are both integuments lacking as in some Santalales). At any rate, the angiospermous ovule was originally bitegmic, and the phylogenetic inquiry may start from the assumption that the Angiosperms descended from forms with two integuments.

The question arises whether the second ovular cover is homologous in all these spermatophytic groups, *i.e.*, in the Coniferophytina, in the seed ferns (including glossopteridalean forms) and in the Higher Cycadophytina (including the Angiosperms). I believe that this is indeed the case insofar as the common early origin is concerned, but in the respective derived groups the second enveloping organ evolved in more or less different ways, becoming adapted to various functions as we have seen and showing a considerable morphological diversity. It was only in the precursors of the cycadopsid groups, the seed ferns, that the third covering organ of the ovule, the cupule, originated. This outer enveloping structure must have developed from 'extraneous' elements in the vicinity of the ovules.

From the accumulated palaeobotanic evidence (LONG 1960, BECK 1962, CAMP and HUBBARD 1963b), one can safely conclude that sterile axial portions (including perhaps some sterilised sporangio- or synangiophores) of the fertile region of a 'mixed' progymnospermous 'frond' became laminose (cladodic) as the result of a process known as 'webbing'. In the Devonian Aneurophytales, the epinastic tips of the rhachis of a fertile pinnule already formed a kind of protection for the aggregated sporangia. The subsequent development, in the form of a branched system of axes, the 'skeleton' of a shallow cupule, is evident from the morphology of the ovuliferous region of such forms as, *e.g.*, the Lower Carboniferous *Eurystoma* (LONG 1960). Although the fossil records indicate cylindric axes as the constituting elements of the protocupules, this does not altogether preclude the participation of some of the 'sterile pinnules' of a compound progymnospermous frond. More fundamental in this connection is the question whether the resulting cupular organ is the homologue of a single 'fertile pinnule' or of several such pinnules. The comparative morphology of the fertile regions of Progymnosperms, Noeggerathiales, seed ferns, *Marsilea*, Salviniales, etc., suggests that the cupule in its attachment to a rhachis bearing coaxial sterile elements corresponds with a sterile progymnospermous pinnule and, together with the ovules it encases, apparently represents at least a whole cladodically transformed fertile pinnule. This means that originally each cupule contains at least all the megasporangia of a fertile pinnule. In Noeggerathiales and glossopterid seed ferns (*Marsilea!*), pleiosporangiate cupules are indeed of frequent occurrence, and, also, in certain ancient Euramerican seed ferns the cupule contained two or more ovules (*e.g.*, the cupules described as *Gnetopsis* and *Calathospermum*). The more advanced seed ferns of lyginopterid and medullosan affinity (including the Salviniales) bear cupules containing a single ovule and one can visualise their possible semophyletic development from the pluriovulate cupules by the secondary oligomerisation of the number of ovules, but a direct phylogenetic origin from a cupule containing a solitary synangium cannot be precluded. Only additional palaeobotanic records can eventually decide whether all mono-ovulate cupules developed secondarily by oligomerisation, or two alternative (pluri- and mono-ovulate) types of cupules evolved simultaneously. However, such minor complications have no bearing on the morphological and phylogenetic interpretation of integuments, cupules and ovules. It is not of key significance whether all the megasporangia of a fertile pinnule, later to become enclosed in a cupule, fused to form a single synangium subsequently changing into a solitary ovule, or whether they fused in smaller groups to evolve into two or more ovules. The cupules of the glossopteridalean seed ferns were at any rate consistently pleiosporangiate and, in some of the more derived forms, such as the Caytoniales, still contained several ovules. The most distinct evolutionary trend among the Cycadopsids was the progressive oligomerisation of the number of ovules, rarely to two (in the only surviving protocycadopsid group, the Cycadales), more often to one (in Corystospermaceae, Peltaspermaceae and all derived, *i.e.*, Bennettitalean, chlamydospermous and angiospermous taxa generally). However, the occurrence of cupule homologues containing two or more ovules in a few Angiosperms cannot be precluded (MEEUSE 1964b).

Previous interpretations of the Caytonialean cupules as primitive 'carpels' (THOMAS) or as the 'pinnules' of a 'sporophyll' can safely be disregarded. As has been shown in Chapter 11, the organ bearing the cupules had already become a distinct gynoclad, an axis bearing stalked cupulate gynosynangia or their homologues (chlamydote ovules). The corresponding radially symmetric female reproductive organ of the Nilssoniales (*Beania*) is of course also a gynoclad and the biovulate appendages of the axis of this 'strobiloid' organ are stalked cupular organs in which only two ovules develop. The recurved broad tip represents the much reduced cupule which does not envelop the ovules. The female reproductive organs ('strobili' or 'cones') of the recent Cycadales are apparently directly derived from the strobiloid gynoclads (co-axial cupules) of the Nilssoniales and are also gynoclads, those of the genus *Cycas* being secondarily modified by the merging of the gynoclad with its supporting bract (pseudo-phyllospory—see Chapter 13) and exhibiting a most peculiar morphological structure. The transformation of the cupule into a flat laminose element has become almost complete in the living cycads, probably because this organ became 'redundant', the protection of the ovules being achieved by the close packing of the ovuliferous appendages, which had also become broader, into a structure superficially resembling a coniferous cone, and the dispersal of the seeds by the fleshy sarcotesta (= outer integument). This is, incidentally, an excellent example of a convergence (an analogy) of non-homologous organs. The biaxial female 'cones' of the cone-bearing Coniferophytina also consist of more or less scaly, densely packed organs helically arranged around a central axis, and the ovules are also protected by the cone-scales surrounding them. Significantly, strongly developed ovular coats ('aril', 'epimatium', or 'sarcotesta') occur in the Coniferophytina chiefly in zoochorous groups (Taxales, Podocarpaceae, *Juniperus*).

In another protocycadopsid group, the Corystospermaceae, a tendency towards extreme oligomerisation of the ovules contained in a cupule can be observed. A striking feature is the presence in these forms of a beak-like extension of the integument of the solitary ovule protruding from the cupule, a condition which is strongly reminiscent of the tubular extension ('tubillus', in CROIZAT's terminology) of the inner integument in cycadopsid groups of Bennettitalean affinities, so well developed in the Chlamydospermae and still recognisable in some traditionally angiospermous taxa. I am of the opinion that the Corystospermaceae, or at least closely related forms with a similar ovular morphology, are the prototypes of at least some Bennettitalean-chlamydospermous groups and indirectly of several major angiospermous lineages. The recognised male reproductive organ of the Corystospermaceae, represented by the form genus *Pteruchus* (TOWNROW 1962) is more reminiscent of the male 'cones' of the Nilssoniales (*Androstrobus*) and Cycadales—which are undoubtedly homologous—than of the stamens of the Angiosperms, but this is not a sufficient reason to reject the obvious semophyletic relationships of the female organs as indicative of a phylogenetic relationship (see also Chapter 17).

The main semophyletic trends in the long orthogenesis of the ovule from the Progymnosperms through glossopteridalean seed ferns to the Mesozoic Cycadopsids were rather restricted, and the salient morphological changes consisted of the development of the two integuments and the cupule and the subsequent oligomerisation of the number of ovules per cupule. There must also have been some reductions in the female gametophyte, but, judging by the conditions in the Cycadales and even in such advanced forms as *Ephedra*, it is more than likely that the Protocycadopsids still had well-developed archegonia and that fertilisation took place by means of zoidiogamy, so that they were not so very much advanced in these respects. The only 'successful' lines that emerged from these Protocycadopsids are apparently the ones in which the number of ovules per cupule became very much reduced and the inner integument acquired a long tegumentary beak.

The acquisition of the third enveloping organ, the cupule, was perhaps the most important evolutionary trend in the Cycadopsids. This third protective layer underwent many semophyletic changes and became adapted to various functions. In the Cycadeoidales the ovules were surrounded by a number of interovular scales forming a tight armour through which only the micropyles protruded, in other groups (Pentoxylales, Chlamydospermae) the bitegmic ovule is enveloped by a fleshy coat (a 'chlamys') and in numerous Angiosperms a homologue of the cupule is present in the form of a true aril.

The origin and nature of the chlamys has been the subject of several discussions, but most of the relevant suggestions are based on false homologies resulting from erroneous and often Angiosperm-centred interpretations. There has been a most confusing interpretation of gnetalian chlamydote ovules as 'female flowers' and of their protective coats as 'perianths' or even 'sporophylls', which need not be discussed. There is cogent evidence, apart from purely typological considerations, that the cycadopsid cupule (*i.e.*, the chlamydote ovule) is the prototype of certain angiospermous reproductive structures (CROIZAT 1947, 1960, MEEUSE 1963b, 1964a, 1964b; see Fig. 11). These angiospermous ovuliferous structures are conventionally known as the 'pistils' ('ovaries'), or sometimes as the 'female flowers' of such forms as Juglandales, Urticales, Piperales and Cyperales, and include many of the traditionally 'pseudo-monomerous' ovaries. The 'female flowers' of the angiospermous taxa mentioned were supposed, on the basis of the Ranalian floral theory, to be very much 'reduced' and to consist chiefly of an equally reduced carpellate gynoecium. If the current interpretation of 'pseudo-monomery' is accepted, the resemblance between the chlamydote gnetalean ovule and the mono-ovulate ovaries (and certain so-called 'female flowers') must be a convergence, an analogy, and indeed many traditionalists consider the

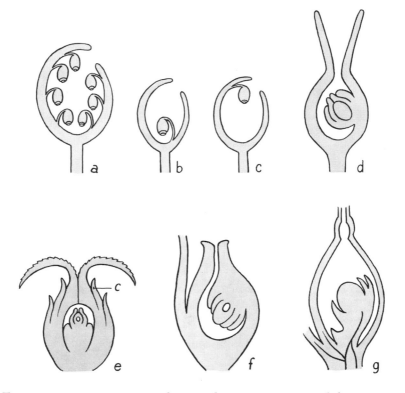

Figure 11. Putative origin of a pseudangiospermous pistil from a pteridospermous cupule: *a*—A primitive protocycadopsid cupule not much advanced beyond the pteridospermous level of organisation (ovules drawn as if pendulous). *b,c*—Cupules after oligomerisation of the number of ovules to a single one, the remaining ovule sub-basically and subapically attached, respectively. *d* to *g*—Pseudangiospermous pistils of *Cannabis* (*d*), *Juglans* (*e*, cupule homologue indicated by letter *c*), *Sarcandra* (*f*) and *Scirpus* (*g*). All figures are in diagrammatic longitudinal section. The tubular extension of the inner integument frequently observed in proto- and hemi-angiosperms is not indicated.

Chlamydospermae to be an independent, highly evolved terminal gymnospermous group not closely related to the Angiosperms, because—and this is the crux of the matter—the gnetalean ovule is not easily compatible with the classical tenet of the leaf-borne ovules. This is a typical example of how 'established' concepts stand in the way of progress, considering that not only the gap between Gymnosperms and Flowering Plants is bridged if the phylogenetic connection between forms with chlamydote 'gymnospermous' ovules and certain groups of the Angiosperms with mono-ovulate gynoecia is accepted, but also that the gradual transference of the pollen-receiving function from the inner integument to the second integument or the cupule homologue (chlamys) demonstrates the continuous transition of gymnospermy into one of the forms of angiospermy or pseudo-angiospermy (see MEEUSE 1964b; also Chapter 12). The origin of the Flowering Plants, since DARWIN's time considered to be 'an abominable mystery', thus resolves itself into a search for their ancestors among the already known chlamydospermous or Bennettitalean groups, which would at once terminate the flood of pseudo-phylogenetic speculations and rid us of several hypothetical protangiosperms. I am satisfied that there are fairly close relationships between Pandanales and Pentoxylales (MEEUSE 1961), and that Piperales and Chlamydospermae must have had proximity of origin. These conclusions in their turn affect a number of morphological and taxonomic inquiries, such as the interpretation of other types of gynoecia (see the following chapter), which not only involves the corner-stone of the classical (Ranalian) floral theory, the 'sporophyll' or foliar carpel concept, but, as we shall see, also the entire floral morphology and the concept 'flower' itself. The phylogenetic assessment of the floral features also necessitates the taxonomic re-evaluation of the revised comparative morphological and phylogenetic interpretations in the classification of the Angiosperms.

The suggestion of the phylogenetic identity of certain angiospermous 'pistils' with gymnospermous (chlamydote) 'ovules' is not new and, if we disregard the muddled semantics and confusing connotations of the traditional Angiosperm-centred approach, essentially the basis of the pseudanthium theory of DELPINO-WETTSTEIN and all its variants (KARSTEN 1918, NEUMAYER 1924, HAGERUP 1934–38, EMBERGER 1944, etc.). One of the stumbling blocks that its protagonists encountered was the vagueness of their own interpretation of the ovuliferous organs of *Ephedra* and *Gnetum* as 'female flowers' with a 'perianth' (and the corresponding male organs as 'male flowers'), and the ensuing confusion of 'ovaries', 'bracts', 'sporophylls' and 'carpels' with a simple cupulate ovule or one of its protective coverings, whereas the essential fact that the cupule (the primitive pistil or so-called female flower) is basically only one out of several co-axial elements belonging to the same gonoclad was overlooked (except by NEUMAYER who, nevertheless, went astray, and by FAGERLIND [1949] who also failed to carry this to its logical conclusion). The high degree of oligomerisation of the genitalia of the recent Chlamydospermae was not sufficiently taken into account. The oligomerisation of the number of cupules is of such frequent occurrence that in several lineages a gynoclad became reduced to a single ovulate cupule, so that sometimes the bract subtending the reduced gynoclad is found in close proximity to the sole surviving cupulate ovule and this foliar element was identified as the bract subtending the 'female flower'. In several instances the cupule became dissected and appears as a 'whorl' of bract-like elements which suggest a 'perianth' surrounding the 'female flower', in fact the bitegmic 'naked' ovule (cf. *Ephedra* and the Juglandales). Such complicated situations were not properly understood, although they all follow logically from the essential homology of a chlamydote (or arillate) ovule and an 'ovary' (a 'pistil') also consisting of a nucellus (the modified megasporangium), two integuments and a third protective layer which can only represent the chlamys (cupule). There are no complicated floral theories, no bracts, perianth lobes or carpels involved in this straightforward semophyletic relation. I expect that this last conclusion especially, an 'ovary' without foliar carpels, will be 'unacceptable' to die-hards of the Old School—even H. H. Thomas (1931), undoubtedly a modernist, tried to relate the caytoniaceous cupule with a Ranalian *carpel!*—so that, in anticipation of possible objections, I submit the following (and, in my opinion, most decisive) arguments, which cannot easily be explained away in terms of the Classical Morphology.

In the first place (see Fig. 12), a connection between the inner integument and the apex of the gynoecium in angiospermous taxa (*Canacomyrica, Engelhardia spicata, Leukosyce*) is manifestly the homologue of the tubular extension (the micropylar tube or tubillus) of the inner integument of the Chlamydospermae and other groups of Bennettitalean affinities. Significantly, the hollow tubillus of *Canacomyrica* still forms an open connection ('stylar canal') between the exostomium of the micropyle and the nucellus.

In *Sarcandra*, it is the outer integument that sometimes protrudes beyond the gynoecium, which presumably reflects the transference of the pollen-receiving function from the inner integument to the second or, eventually, the third protective ovular coat. The gynoecial morphology of *Engelhardia* is very illuminating in this connection. A longitudinal section of the so-called 'female flower shows a peculiar arrangement of the three protective organs which formerly was frequently interpreted as indicative of a perigynous or hypogynous condition of the ovary. The so-called 'perianth' or 'calyx' is the homologue of the cupule (which does not encase the ovule completely!) and the 'ovary wall' represents the outer integument. The micropylar tube of the inner integument continues imperceptibly into the inner (upper) surface layer of the stigmatic extensions of the pistil, which strongly suggests that at least parts of the stigmatic areas were formed as differentiations of the distal portion of the tubillus, in other words, the pollen is caught by an integument, not by

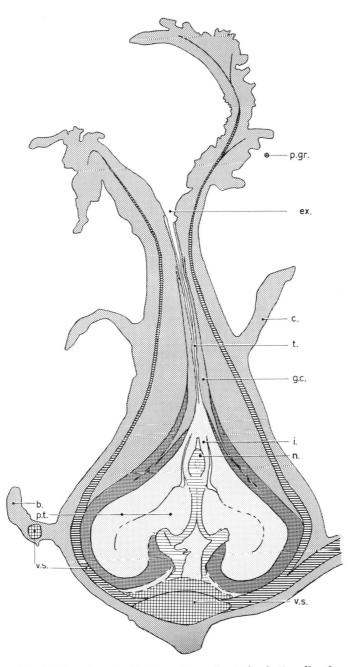

FIGURE 12. Median longitudinal section of pistil of *Engelhardia spicata*, approximately ×30: *p.gr.* = pollen grain (indicating complete stage of anthesis!); *ex* = exostomium of micropylar tube or tubillus (*t*); *c* = lobe of the cupule derivative (chlamys), *i.e.*, of the so-called 'perianth' or 'calyx'; *g.c.* = zone of gum cells; *i* = (inner) integument; *n* = nucellus; *b* = lobe of bract (later forming the large 'wing'); *v.s.* = vascular strands. Compare the microphotograph (Frontispiece) corresponding to this figure (see also MEEUSE and HOUTHUESEN 1964 for additional details).

the cupule or by a part of a carpel. In the gynoecia of the other genera of the Juglandaceae (and in other species of *Engelhardia*), the distal end of the micropylar tube forming the 'stylar canal' and the stigmatic region is phylogenetically (and presumably also histogenetically) severed from the perinucellar part of the inner integument, but the Juglandaceae are technically still 'gymnospermous'. It is interesting to note in this connection that LEROY (1954) most emphatically denies the presence of any indication that the juglandaceous pistil ever contained more than one ovule. In view of the origin of this type of pistil from a cupule, LEROY's statement is, strictly speaking, untenable, but it remains nevertheless highly significant in that he considers, at the angiospermous level, the juglandaceous pistil to be primarily mono-ovulate and not secondarily derived from pluri-ovulate carpels.

The conventional interpretation of the juglandaceous and other forms of cupular gynoecia as pseudo-monomerous pistils of carpellar derivation requires three assumptions: (1) such a monovulate ovary must have been derived from an ovary built up of several carpels; (2) all the carpels but one became sterile; and (3) all the ovules of the remaining fertile carpel but one became reduced. When the ovule is sub-basally attached, *i.e.*, pseudo-terminal (Piperaceae, Juglandaceae, some Urticales, etc.), a fourth supposition must be added, *viz.*, (4) a 'shift' of the solitary ovule from its postulated originally marginal place of insertion to a basal position. Unlike the condition in families in which truly pseudo-mono-merous *carpellate* ovaries occur (Ranunculaceae, Myristicaceae, Rosaceae), there are no intermediate stages linking the so-called 'reduced' ovaries of Piperaceae, Monochlamydeae and Pandanales with the alleged multi-ovulate prototypes. On the contrary, when a related taxon has a pluri-ovulate carpellate type of pistil, there is a gap: they represent different 'levels' because the fundamental semophyletic relation is not between whole conventional 'pistils' or 'ovaries', but between cupulate (arillate) *ovules*, so that, for instance, the traditional 'pistil', actually a transformed cupule, of, *e.g.*, the Piperaceae and Lauraceae, is not homologous with the carpellate type of ovary of, say, the Magnoliales, but only with one of the (arillate) ovules inside such a ranalian follicle. The two types of functional 'pistils' represent divergent evolutionary tendencies resulting in parallel phylogenetic lineages, the one lineage having retained the modified and usually one-ovuled cupule as the functional pollen-receiving and seed-incubating unit, and the other having evolved a different functional gynoccial unit comprising several to many ovules encased in an entirely new form of pistil wall. Almost needless to say, the 'pistils' of cupular derivation represent the phylogenetically older (more primitive) condition, the carpellate ones, as derivatives of whole gynoclads and their subtending bracts (see the next chapter), a more derived form (a higher semophyletic level) of the angiospermous gynoecium. The taxonomic and typological implications of the occurrence of two alternative categories of 'pistils' are self-explanatory.

Another point to consider in connection with the last paragraph is the inconsistency resulting from the 'old' floral morphology when the gynoecia must be declared to be very advanced in a group like the Chloranthaceae, although the evidence (vessel-less primitive secondary xylem, archaic types of pollen grains, continuation of the development of the embryo after the shedding of the seed and before germination) points decidedly to the Chloranthaceae as the most primitive family of the Piperales. Similarly, the pistils of *Freycinetia* must be considered to be the most primitive of all Pandanales, but the general impression is that the ornithophilous and partly chiropterophilous genus *Freycinetia* is the most highly evolved. Heterobathmy (TAKHTAJAN) is undoubtedly an important factor in organic evolution, but surely not to that extent.

Finally, the typological arguments of the classical floral morphology cannot carry much weight when the palaeobotanic records indicate that cycadopsid plants with ovuliferous cupules were common in the Mesozoic, whereas potentially protangiospermous fossils exhibiting the postulated 'sporophylls' are conspicuous only by their absence. The adoption of the homology of chlamydote ovules (modified cupules) with certain traditional 'pistils' opens up the road to so many deductions which dovetail logically and explain several awkward but only apparent inconsistencies so satisfactorily that the postulation of this phylogenetic relation at least provides a sound working hypothesis which is compatible with a great deal of typological, palaeobotanic, anatomical, palynological and phylogenetic evidence. As an additional advantage, the new interpretation brings the morphology of the reproductive organs of the Angiosperms completely into line with that of the Cycadopsids, *i.e.*, of Gymnosperms, and enables the reconstruction of orthogenetic relations linking the Angiosperms with the Upper Devonian Progymnospermopsida, in other words, paves the way for a veritable comparative floral morphology of all Spermatophyta. That the cherished 'sporophyll' concept is thus made an orphan child seems to me only a small sacrifice in proportion to the gain.

An objection that is certain to be raised by protagonists of the classical floral theory is that the solitary ovule of a 'pseudo-monomerous' ovary is not always basally attached (*i.e.*, not terminal on the supporting axis). In many cases that have been studied, the histogenetic differentiation of the ovule proper begins with the formation of a lateral bulge on a lateral primordial outgrowth of the shoot apex and not as a terminal differentiation of the shoot apex itself. The phytomorphologists who defend the 'appendicular' origin of the ovules of such 'ovaries', such as ECKARDT (1937) and EAMES (1961), claim this to be a convincing demonstration of the foliar nature of the organ bearing the ovule. However, the argument is not based on phylogenetic evidence, but on circular reasoning: every lateral primordium of the shoot apex must be the primordium of a phyllome, so that an ovule formed on such a lateral primordium is, accordingly, 'leaf-borne' ('appendicular'); and, conversely, the lateral origin of an ovule ('leaf-borne' by postulation) on a primordium is supposed to be indicative of the foliar nature of the ovule-bearing organ developing from that primordium. In the discussion of pseudo-phyllospory (Chapter 14) I have discussed the fallacy of this deduction when applied to 'fertile' apices. The lateral position of an organ bearing an ovule does not mean that this organ is a phyllome, a 'leaf', even if it is claimed that the histogenetic development is a yardstick of its phylogenetic history (HAECKEL's law of recapitulation), because the phylogeny of the foliar 'carpel' is a moot point. The neomorphological approach indicates that the solitary ovule in the case under discussion represents the only remaining ovule of a reduced cupule.

In other words, all cycadopsid and, accordingly, all angiospermous ovules are fundamentally cupule-borne and not 'leaf-borne', so that the relative positions of the ovule(s), the cupule (or its homologue, the ovary wall) and the central floral axis are phylogenetically determined and can be deduced from the morphology of a suitable archetype. The origin of the cupule from at least a whole progymnospermous 'fertile pinnule' provides the basic pattern of a triaxial structure, the proximal portion of the rhachis of the pinnule (the future funicle) and its continuation in the cupule representing a 'lateral' axis of the rhachis of a 'mixed pinna', and the synangiophores representing subsidiary axes of a higher order which are in their turn 'lateral' in respect of the axis of the pinnule. The synangia and their derivatives, the bitegmic ovules, are accordingly laterally attached on the cupule (the possible occurrence of an occasional single synangium or ovule in a truly terminal position on the rhachis of the 'fertile pinnule' can be disregarded). The progressive oligomerisation of the ovules in the cycadopsid cupule may have resulted in the retention of the most proximal one, but it is equally probable that a more distally situated ovule remained fertile. The place of insertion of the remaining fertile ovule on the cupule must be expected to vary from proximal (sub-basal) to distal (subapical), so that in the 'pseudo-monomerous' pistils of cupular derivation various forms of placentation except a truly basal (i.e., 'terminal') ovule are likely to occur. It is quite clear that the lateral attachment of the ovule does not 'prove' the foliar nature (phyllospory) of the pistil. The histogenetic development of the pistils of cupular derivation is of course perfectly compatible with their suggested phylogenetic origin, because the origin of the 'ovary wall' as a lateral outgrowth of the fertile shoot apex re-

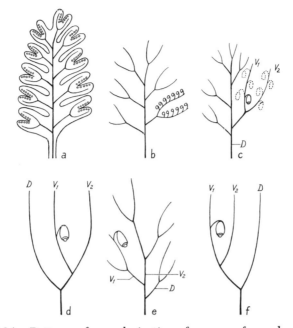

FIGURE 13A. Patterns of vascularisation of organs of cupular derivation. A primitive stage is represented by a marsileaceous type of cupule (diagrammatic): *a*—Vascularisation of the sporocarps of *Marsilea*. *b* and *c*—The oligomerisation and reduction of vascular strands and ovules, leading to *d*, the ultimate stage, currently interpreted as the standard vascular pattern of a 'monocarpellate' structure with a so-called 'dorsal' bundle (*D*, unpaired) and a pair of 'ventral' (or 'lateral') strands (V_1 and V_2), innervating the ovules (in the figure only one ovule is drawn—the condition found in some of the so-called pseudo-monomerous pistils). *e,f*—A similar reduction series leading to a pattern corresponding with that of a 'pseudo-monomerous' ovary, a phylogenetically different main strand representing the so-called 'dorsal' (*D*) and two pairs of strands the 'ventrals' (V_1 and V_2), previously interpreted as a bicarpellate structure.

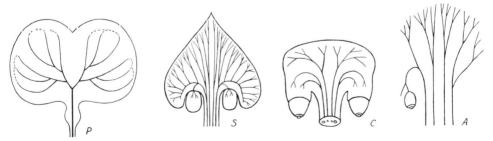

FIGURE 13B. Patterns of vascularisation of the cupules of early cycadopsid forms: *P—Pilularia* type. *S—Stangeria* type. *C—Ceratozamia* type. *A*— Vascular pattern of an angiospermous pistil, interpreted as indicative of a bicarpellate condition.

flects the lateral insertion of the cupule on the supporting axis (the axis of the gynoclad), and the ovule arises as an 'appendage' of the young ovary wall in agreement with the lateral attachment of the ovule to the inner surface of the cupule.

Many phytomorphologists, from VAN TIEGHEM (1875) to EAMES (and his associates, 1931–61) and ECKARDT (1937), thought to have found 'irrefutable' anatomical evidence supporting the conventional interpretation of the so-called pseudo-monomerous gynoecium as a reduced monocarpellate or syncarpous pluricarpellate ovary. The more or less satisfactory correspondence between the actual vascular anatomy of the pistil and a preconceived pattern of a 'dorsal' and (two) associated 'ventrals' (or 'laterals'), reputedly representing the main vascular strands of a foliar 'carpel', enabled them to 'recognise' a number of such 'carpels' in the ovary wall. Reduced to its bare essentials, this interpretation rests upon the assumption that a certain stereotype arrangement of vascular bundles is exclusively indicative of the presence of a specific category of organs (the carpel). I have explained in detail elsewhere (MEEUSE 1964b) that during the semophylesis from cupule to pistil the vascular skeleton of primitive cycadopsid cupules (for which those of *Marsilea* and the Cycadales served as examples) may conceivably evolve into certain structural patterns which agree with the actual vascular anatomy of so-called pseudo-monomerous gynoecia (see Figs. 13A and B), so that the mere presence of the patterns of so-called 'dorsals' and 'ventrals' has no demonstrative force in a discussion of the nature of these pistils. Since only the corroborative and unequivocal evidence from phylogeny, typology and histogenesis is acceptable as relevant in the morphological interpretation of a complex organ with a long and checkered evolutionary history, the inescapable corollary is that the gynoecia ('pistils') of the more primitive Angiosperms are modified cupules (Juglandales, Urticales, Piperales mostly, Cyperales, most probably also Laurales and *Nelumbo*), or aggregates of such transformed cupules (co-axial 'phalanges' in Pandanales, *Centrolepis;* lateral coalescence of pistils belonging to different gynoclads in Arecales, Restionales). A most tempting conjecture is the possible retention of a truly archaic cycadopsid feature, *viz.*, of the bio-ovulate (or pluri-ovulate) condition, in the cupular gynoecia of some angiospermous taxa. I am inclined to accept this as indeed the case in *Calycanthus* and *Casuarina*, perhaps also in some amentiferous orders such as Betulales and Fagales. A re-examination of these taxa in the light of this new interpretation will certainly prove to be rewarding.

A striking feature of the ovule is the diversity of its position in respect of its supporting stalk (the synangiophore or funicle) and of the point of attachment of the stalk. It is an established fact that the different conditions known as orthotropy, anatropy, campylotropy, epitropy, apotropy, etc., are often of constant occurrence in certain taxa and hence provide excellent diagnostic characters. This specificity betrays a phylogenetic basis of the development of the various forms of ovular insertion, so that a tentative analysis of semophyletic changes does not seem out of place.

PLATE IV. *Pandanus tectorius,* female specimens growing in *Barringtonia* formation, 'fruiting.' Southern coast of western Java near Udjung Genteng. (Photo by Dr. C. G. G. J. VAN STEENIS.)

The problem is complicated by the fact that although one has been wont to call tegumented megasporangia 'ovules' and their stalks 'funicles', irrespective of the presence or absence of a cupule (aril), one should distinguish between the bitegmic *ovules* of the Coniferophytina and Lower Cycadopsida (and the Higher Cycadopsida with ecarpellate gynoecia) and the derivatives of one-ovuled *cupules,* the chlamydote or arillate ovules of chlamydospermous, Bennettitalean and many angio-spermous groups. It is clear that the stalk of the ovule, *sensu stricto,* is a synangiophore and the stalk of the arillate ovule a completely different organ, *viz.,* the stalk of the cupule. The semophylesis of the ovule proper is decided by the changes of the relative position of the ovule and its synangiophore in respect of the supporting axis, but the presence of a cupule to a large extent involves the semophylesis of the cupular stalk and the cupule, including the changes in their relative position in respect of the supporting axis (gynoclad) and of the ovule contained in the cupule.

The primitive ovule was erect and also orthotropous in respect of its synangiophore. This condition was retained in many forms, such as seed ferns, Ginkgoales, Taxales, and prevails also in many Cycadopsida if the position of the ovule in relation to its own synangiophore is con-sidered. A common feature is the incurvation of the synangiophore so that the ovule, though technically orthotropous in respect of its stalk, is pendulous. This condition is known from, *e.g.,* Cordaitales and a com-parable position of the 'ovule' is also found in several Angiosperms. In this situation, the nucellus is parallel to the ovular stalk and the micropyle faces the point of attachment of the stalk. During the evolution of the Pinales from a prototype of cordaitalean affinity, the ovular stalk became incorporated in the ovuliferous scale when the ovule became adnate to the scale, so that the intimate fusion of synangiophore and ovule re-sulted in an 'anatropous' condition. In the Higher Cycadopsida, the situation was undoubtedly much more complicated, but there is no reason to assume that in a primitive cupule containing several ovules an early re-orientation of the ovules in relation to their individual synangiophores was of common occurrence, because several derived groups still have orthotropous erect ovules. However, in some cases the ovules must have become pendulous in respect of the axis of the cupule if their synangio-phores were, or became, so long and flexible that their micropyles were not necessarily directed towards the orifice of the cupule. This again had certain consequences in connection with the fertilisation process, primarily with the catching of the microspores, which in the early and still zoidiogamous Cycadopsids had to reach the exostomium of the ovular micropyle. The orientation of the ovule was thus partly deter-mined by certain ecological factors, so that in the Corystospermaceae the ovule was already provided with a tegumentary tube which was bent in an upward direction, apparently because a pollination droplet formed on the micropyle is more persistent and hence more effective if sup-ported from below than if borne in a hanging position. The consequences of such a development can be deduced from an abnormal (most probably atavistic) gynoecial condition in *Sarcandra* (see Fig. 14). The pistil (cupule!) is erect, but the ovule is more or less pendulous and the tegumentary tube is recurved towards the stigmatic area of the pistil. In this case, the semophyletic history can be read with a high degree of probability (see Fig. 14): in the most primitive condition, the ovule was erect on its synangiophore, *i.e.*, perpendicular to the lateral side of the cupule. It became pendulous and developed a recurved micropylar tube. When it subsequently became completely encased by the cupule, it retained its position but the redundant micropylar extension (no longer required after the advent of siphonogamy) disappeared. In the related Piperaceae (and in the Juglandales), the retained ovule is sub-basal and remained erect in respect of the cupule. The tubillus (still present in

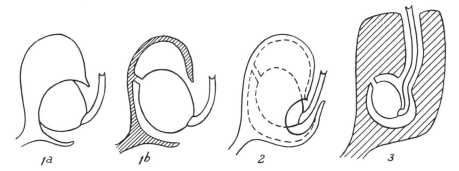

FIGURE 14. Suggested origin of a chlamydospermous female reproductive organ (or an angiospermous gynoecium) from a protocycadopsid precursor: *1a*—Lateral view of monovulate cupule of a corystospermaceous form, the ovule with a tubillus, in longitudinal section represented in *1b*. *2*—The sterile cupule wall has almost enclosed the ovule. *3*—The gynoecial wall has enclosed the ovule completely, the communication between the ovule and the outside world being maintained by the tubillus (compare Fig. 11*f*).

Engelhardia spicata) was straight and erect, so that after the partial reduction of the tegumentary tube a 'pistil' with a sub-basal, erect and orthotropous ovule emerged. The axis of the ovule could apparently re-orientate itself independently of the longitudinal axis of the cupule, but alterations of the shape of the cupule and changes in its position in respect of its stalk may conceivably have exerted some morphogenetic effect on the ovule. A fusion of the pendulous ovule with its synangiophore resulted in the formation of anatropous ovules, and there are no reasons to preclude the possibility of a re-orientation of a pendulous orthotropous ovule on its stalk so that the micropyle faced upwards instead of downwards previous to the advent of anatropy. There are so many possible alternatives that a considerable divergence in the ovular morphology developed which was certainly increased when one-ovuled cupules (*i.e.*, arillate ovules) became enclosed in a carpellate pistil.

Additional re-orientations, partly induced by the morphogenetic effect of neighbouring elements and partly by ecological requirements, in conjunction with the varying relative position of the synangiophore in respect of the stalk of the cupule, again caused changes in the ovular morphology and also in the orientation and insertion (placentation) of the ovules.

The characteristics of the ovule thus reflect a long and complicated phylogenetic history, so that their constancy in smaller taxonomic categories of the size of families or orders, *i.e.*, their considerable diagnostic value, is not all surprising.

16
Phylogeny of the Megasporangium:
II. The Carpellate Pistil

In its appendicular theory of the flower, the typological phytomorphology postulated the homology of the carpels, in other words, the
Angiosperms were supposed to possess only one fundamental category of
female organs, *viz.*, the carpel. Accordingly, all pistils were interpreted as
structures consisting of, or derived from, one to many 'carpels', and even
if their morphological features were rather unusual (free central placentation, solitary basal ovules) they were all squeezed into this pattern. It
is not without significance that the original appendicular hypothesis was
repeatedly emended, because the vascular anatomy of the gynoecia required the postulation of a primitive condition of leaf-gaps and leaf-
traces (see, *e.g.*, EAMES 1961) and this, in conjunction with incongruous
forms of placentation, necessitated the assumption of pinnate, palmate or
peltate carpels to fit the vascular pattern and other morphological features such as free central and basal placentation. MELVILLE (1962) has
pointed out that there are other inconsistencies which cannot satisfactorily be explained in terms of the classical sporophyll doctrine, such as
the fusion of the 'ventral' bundles of *adjacent* follicles to form a placenta,
and rejects the appendicular theory of the carpel almost categorically.

The phylogeny of the ovule, discussed in the previous chapter, is a
sequence of semophyletic processes, not only involving the original
'core', the megasporangium and its derivatives including the gametophyte,
but also the successively acquired 'protective covers', the two integuments
and, in the Pteridosperms and Cycadopsids, a third envelope, the cupule,
'chlamys' or true aril. This three-layered ovular coat is still found in
many traditional Angiosperms as we have seen, the chlamys in these
taxa forming the wall of the gynoecium or a large portion thereof. The
classical phytomorphology did not consider this possibility; and even
R. VON WETTSTEIN and other adherents of the pseudanthium theory, although they built up a floral morphology starting from a chlamydospermous prototype, were too much under the ban of classical concepts to
recognise the exact semophyletic relations. Their reasoning was still so

Angiosperm-centred that either the cupulate chlamydospermous ovule was called a 'female flower' and its chlamys was supposed to be a 'perianth' (WETTSTEIN), or the angiospermous ovary was thought to be a derivative of one to several gnetalian 'female flowers' and a number of enveloping 'bracts' (KARSTEN, FAGERLIND 1946). The direct semophyletic connection between the chlamydote gnetalian ovule and the 'one-ovuled pistil' of several angiospermous groups was not clearly understood or hopelessly confused by too complicated an interpretation (NEUMAYER, HAGERUP). THOMAS (1931) also attempted to relate the gynoecia of the Angiosperms with those of cycadopsid Gymnosperms, but although he was undoubtedly on the right track when he suggested that the Caytoniales are potential Protangiosperms, he inadvertently confused lines and levels by comparing the Ranalian *follicle* with the caytonealean *cupule* (the *cupule,* as we have seen, is retained in the carpellate angiospermous ovaries as the true *aril*).

The inquiry into the phylogenetic origin of other types of gynoecia must start from the homology of all ovules, including their individual investing covers, and this means that one must start searching for the third protective coat in all Angiosperms.

Considering that the 'stalk' of the ovule, the synangiophore, became incorporated in the funicle, it is clear that the true aril, borne on the funicle below (outside) the integuments, represents the chlamys or cupule. The colour, the consistency and the function of the aril is often comparable to that of the fleshy cupule or chlamys (*i.e.,* to the often fleshy ovary wall in angiospermous taxa with pistils of cupular derivation: Pandanaceae, some Cyperaceae, several Monochlamydeae and Piperales), reflecting its early adaptation to zoochory. The presence of several arillate ovules in a common containing structure (the wall of the pistil) indicates that the encasing ovary wall is of 'extraneous' origin and not a product of the ovular coats, all three individual enveloping layers of each ovule already being accounted for. The obvious corollary is that a group of ovules is closely associated with a sterile organ or with several such organs. The fertile regions of the Cycadopsida are fundamentally anthocorms, the female portion of which essentially consists of a number of cupuliferous axes (gynoclads) each subtended by a bract. The carpellate pistil must semophyletically have originated from a number of such gonoclad-bract units, the fertile axes contributing the placentae with their (arillate) ovules and the bracts providing the valvular parts of the walls and the septa. The structure is supported by the central axis which, at least in ovaries compounded of several gonoclad-bract units, often forms the central core of the pistil (thus contributing largely to the part called receptacle or torus) and is sometimes developed as a discrete gynophore or androgynophore. The pistil is crowned by the pollen-receiving stigmatic areas, which are frequently connected with the ovary by an attenuate and almost invariably apical extension, the style, or by several such styles. The styles and stigmata are often complex structures which may contain elements of the ovular coverings, the gynoclad(s), the bract(s) or stegophyll(s), and the central axis of the anthocorm (the common floral axis) as will be demonstrated presently.

The ovuliferous axes initially bore several to many cupules, so that there was a primary association of *one* lateral bract-like organ with *numerous* ovuliferous elements based on a common axis (or placenta). This is where protagonists of the pseudanthium hypothesis went astray, their fundamental unit being an axis with a single 'terminal' ovule, which is indeed the basic form of gynoecium in several groups as we have seen, but cannot possibly be the prototype of the pluri-ovulate pistils of, *e.g.*, the Polycarpicae, Centrospermae, many other choripetalous and all sympetalous Dicotyledons, Helobiae, Arales, Scitamineae, Liliales and their respective derivatives. The number of cupules (arillate ovules) of a 'naked' gynoclad may be reduced to a single one, as frequently happened in the phylogeny of monocotyledonous groups (*e.g.*, of the Cyperaceae), but the oligomerisation of the number of cupules may also have taken place after the gynoclad had been incorporated in a carpellate pistil. The comparative morphology of the gynoecia of the Ranunculaceae indicates such a 'secondary' reduction to a one-ovuled condition in the Anemoneae. On the other hand, in taxa which have pseudo-angiospermous pistils of the 'chlamydote ovule' type, the funicles are sometimes laterally attached, as we have seen in the previous chapter (*e.g.*, in *Cannabis, Centrolepis,* and *Typha*), and it may become very difficult to distinguish between the one-ovuled truly 'pseudo-monomerous' carpellate ovary and the gynoecium of cupular derivation (MEEUSE 1964a). In either case, the outer layer (the pistil wall, *i.e.*, either a derivative of a sterile bract or several such bracts, or the homologue of the chlamys) may have evolved in different directions as a result of adaptive evolution and become dry, hard, fibrous or spongy after maturation, so that one would not so readily identify a dry, fluffy, fibrous or hard layer with a chlamys (an aril!). It is possible to distinguish between the two categories of mono-ovulate gynoecia by using the following criteria:

1. The presence of an aril around the ovule inside the pistil wall (Myristicaceae, which possess, accordingly, carpellate ovaries).
2. The comparative morphology of a group indicating a 'reduction series' by exhibiting the progressive oligomerisation of the number of ovules (Ranunculaceae-Anemoneae, Rosaceae, some Leguminosae, several 'centrospermous' groups such as Polygonaceae, Chenopodiaceae, many Amaranthaceae, etc., and various advanced sympetalous taxa such as Valerianaceae and Compositae).

3. The mode of dehiscence: the separation along a longitudinal suture usually indicates a follicular carpellate ovary (*Myristica*).
4. Indications from the vascular anatomy, the histogenetic development, and androecial morphology ('isomery'), or other features that the ovary wall consists of two or more fused elements point to a pseudo-monomerous condition (*Polygonum*).

Nevertheless, the evidence may not be conclusive, and for the time being, I cannot decide to which category the pistils of, *e.g.*, *Nelumbo*, Lauraceae and Monimiaceae must be referred (I think they are ecarpellate). At any rate, the pseudo-monomerous carpellate ovaries represent only a special case, and their initial semophyletic development was not essentially different from that of corresponding pluri-ovulate types.

The phylogenetic prototype of the carpellate pistils consists of a unit of a gynoclad and its associated bract or of several such units, but the development of the closed ovaries from such dual structures must have proceeded along two more or less different evolutionary pathways. In certain cases, the elements of the units retained their individuality up to a point and the placental region remained rather clearly separated from the sterile portions, which is reflected in the histogenesis of such ovaries, the placentae developing independently of and frequently even before the bracts ('pseudo-carpels'), often in an axillary position in respect of these bracts (see, *e.g.*, PANKOW 1962). This is the normal case in the Centrospermae (MOELIONO, unpublished), but there are several minor variants. The placental axes may be fused into a central column, with or without the participation of the main floral axes, or they are redundant (as in Portulacaceae), or reduced to a single basal ovule (Basellaceae, many Amaranthaceae, Chenopodiaceae). The bracts forming the pistil wall usually meet at the apex, thus forming a dome-shaped structure over the fertile axes, and often joining up with the central floral axis distally of the ovules as in Caryophyllaceae. All these closed 'centrospermous' pistils developed semophyletically by the coalescence and concrescence of a 'whorl' or tier of bracts ('pseudo-carpels') into a barrel-shaped structure which was originally open at the distal end and did not take part in the pollination process, the pollen grains being caught by the individual ovules or by apical extensions of the placental regions to be discussed presently. Eventually the ovary wall became closed at the apex. It is a characteristic feature of these ovaries that the ovary wall is always constituted by several bracts, even if the fertile axes are reduced to a single stalked ovule, which also applies to the phenetically very similar but presumably homoplastically evolved gynoecia of the Primulales.

Another extreme case is the consistent association of a single fertile axis and its subtending bract as found in the follicular pistils of the apocarpous Polycarpicae, Leguminosae and a few smaller groups. As has been shown in Chapter 11, these organs, in the traditional appendicular theory interpreted as (foliar) carpels, 'eucarpels', 'eusporophylls', etc., and indeed giving the impression of being entities behaving like 'single' structures, are actually of dual nature, amalgamations of gynoclad and bract, which I have called pseudo-sporophylls for historical reasons. The term 'carpel' can be retained, provided the meaning is restricted to a derivative of a unit consisting of a gynoclad and its associated bract. 'Carpels' thus defined can, accordingly, also be recognised in other categories of gynoecia, including the syncarpous (and paracarpous) ones which contain several gonoclads and their corresponding bracts.

Apart from the 'centrospermous' type of ovary, there is another group of gynoecia which are apparently also semophyletically derived from a number of gynoclad-bract units arranged in a whorl, but which have other forms of placentation. The rather fundamental conclusion following from this postulated phylogenetic development is that the conventional (typological) derivation of all coenocarpous ovaries from a group of apocarpous (monocarpellate) Ranalian follicles is decidedly erroneous. Syncarpous ovaries that originated from a secondary fusion of apocarpous elements are not altogether non-existent, but surprisingly rare and perhaps restricted to the Polycarpicae (*e.g.*, several Annonaceae, *Nigella* in the Ranunculaceae). All other cases originated as a primary aggregation of several gynoclad-bract units which simultaneously coalesced, *sometimes before each gynoclad-bract unit became an individually closed structure and at any rate before the whole gynoecial structure became a closed pistil*, so that the pistils of many angiospermous taxa can be said to have originated at once as coenocarpous complexes. The phylogenetic implications are far-reaching: the most timid claim one can make is that, if the Magnoliales (or all Polycarpicae with apocarpous monocarpellate follicles for that matter) are related to other major angiospermous taxa, this relationship can only be through propinquity of descent from a common ancestral *ecarpellate* group with 'naked' gynoclads. Another reasonable assumption is that the adaptive evolution of the Angiosperms emanating from the 'protection of the ovules' would act on every whorled aggregate of bracteated gynoclads and transform such aggregates polyrheithrically into closed ovaries; in other words, the typology of such coenocarpous gynoecia is not necessarily indicative of their true semophyletic relations and, consequently, of a close phylogenetic origin of two taxa with a phenetically very similar gynoecial morphology. This argument is considerably strengthened by the repeated occurrence of carpellate ovaries among homogeneous taxonomic groups which also comprise ecarpellate gynoecia of the 'chlamydote ovule type'. The semophyletic sequence can only be read in one direction, the carpellate ovaries being derived from ecarpellate gynoclad-bract units, so that within such a taxonomic group the carpellate forms must always have evolved from ecarpellate prototypes related to the extant taxa which retained the more primitive gynoecial morphology. The semophyletically younger carpellate pistils must in such groups have developed as parallelisms. Illustrative examples are the carpellate Saururaceae among the Piperales, the carpellate Salicaceae among the Amentiflorae, perhaps the genus *Freycinetia* among the Pandanaceae. The carpellate ovaries of these taxa are not directly semophyletically related, nor are they phylogenetically derived from the carpellate gynoecia of any other group. A typology of carpellate pistils is possible, provided it is based on an ecarpellate prototype. The gynoecial morphology is apparently *not* such a very reliable yardstick of phylogenetic interrelations, and this widens the scope of taxonomic research considerably.

The two extreme cases of association of gynoclad and bract are connected by intermediate conditions. The association is a fairly loose one in primarily coenocarpous ovaries with central or 'axile' placentation, the septa usually developing centripetally and only secondarily establishing contact with the placental regions (and sometimes 'bisecting' a placental zone derived from a single gynoclad [see also MELVILLE 1962]). In coenocarpous ovaries with parietal or laminal placentation the ovuliferous portions of the gynoclads became associated with the bracts, but not always with the central floral axis, and the constituting carpels may approach the extreme pseudo-phyllospory of the apocarpous mono-carpellate Ranalian follicles. The various forms of placentation are likewise *alternative* conditions, and there is no reason to consider one type to be more primitive than another, let alone to derive all forms of placentation from the same (*i.e.*, the postulated marginal) type. Some pistils must have had parietal, laminal or axile placentation before the carpels had formed an entirely closed ovary, so that laminal placentation in the gynoecia of, *e.g.*, Nymphaeaceae *s.s.* and *Butomus* is presumably as much an original feature characteristic of these taxa as the axile insertion of the ovules in Dilleniales and Guttiferae and the parietal placentation in the 'Parietales' (Cistiflorae).

The question arises as to whether intermediate stages of the semophyletic development of the carpellate ovary from gynoclads and bracts can still be found among the recent Angiosperms. The incompletely closed carpels of some Winteraceae (*Drimys* sect. *Tassmania*) and of *Degeneria*, although neatly demonstrating the development of the stigmatic crest as the origin of a carpellary stigma and style, do not show the dual nature of the pseudo-sporophyll. The carpels of the Dilleniaceae are sometimes incompletely closed at the distal end. The traditional tenet that coenocarpous gynoecia are derived from fused apocarpous follicles is not consistent with the morphology of such dilleniaceous gynoecia, which are strongly suggestive of a whorl of conduplicate bracts more or less fused at the base and each of which subtend a placenta which is adnate to the floral axis.

Much more convincing, to my mind, is the gynoecial morphology of the Centrolepidaceae. The floral morphology of this group has recently been discussed by HAMANN (1962), who pointed out that the arrangement of the male and female organs in *e.g., Centrolepis* is indicative of a 'pseudanthial' complex structure rather than of a 'flower'.

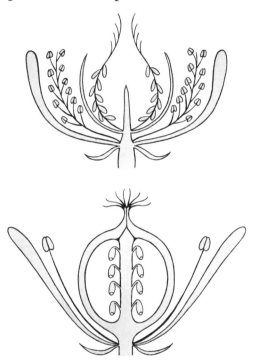

FIGURE 15. Tentative derivation of the origin of the gynoecium of the liliiflorous type from an aggregate (whorl) of bracteated gynoecial structures (resembling those of *Centrolepis*) by the coalescence of the ovuliferous gonoclads and the closure of the bracts.

In my opinion, the aggregate of one-ovuled elements on a common axis subtended by a bract must be interpreted as a 'phalanx', *i.e.*, as a gynoclad with coalesced chlamydote ovules. The ovules either have an individual 'stylar' extension, or they have a common 'style'. This whole structure is strongly reminiscent of a placenta and the longitudinal fusion of three such gynoclads, followed by the envelopment of the aggregate by the bracts, would lead to a carpellate ovary of the liliaceous type (see Fig. 15). The merging of such a gynoclad with its bract would produce a pseudo-sporophyll and, if a whorl of such pseudo-sporophylls developed simultaneously, a coenocarpous ovary with laminal or parietal placentae would be formed. The Centrolepidaceae provide the clue to the semophyletic development of various categories of monocotyledonous carpellate ovaries (liliaceous-juncaceous, but also, *e.g.*, xyridaceous) as well as showing a tendency towards the oligomerisation of the number of ovules in some genera, thus demonstrating the phylogenetic origin of the non-carpellate gynoecia of Arecaceae, Cyperaceae, Restionaceae, Gramineae, etc. The presence of stylar extensions in *Centrolepis* is very significant because it indicates that either the individual ovules or the placental regions (gynoclads) already possessed pollen-catching structures before they became incorporated in a closed carpellate ovary. The vital process of pollination, as the initial phase of fertilisation, was not at all changed or affected before the closing phases of the encasing of the placentae by the bracts of the gynoclads, but ultimately the enveloping bracts must have come in close contact with the ovular or placental style or styles. In most instances, the distal continuation of the placental vascular strands into the style(s) and stigma(ta) at least indicates the importance of the original vascular connection between the placental region of the gynoclad and the stigmatic pollen-catching region, a point also emphasised by MELVILLE (1962). It is conceivable that, although the sterile carpellary wall often participates in the formation of styles and stigmata (perhaps as a glove-like cover of the original tegumentary or placental extensions) and contributes to their vascularisation, the old supply route from the placenta is maintained to provide water, sugars and other physiologically active substances necessary for the germination of the pollen grains and the growth of the pollen tubes. MELVILLE (1962) suggests that the closure of the carpellate gynoecium was already initiated when the fertilisation was still established by zoidiogamy, so that a continuous film of liquid must have been present on the inner surface of the carpel between the stigma and the ovules, but this is most unlikely. Siphonogamy (and double fertilisation) must have become established fairly early in the Mesozoic, at any rate long before complete 'angiospermy' was attained: even such archaic relict forms as the Chlamydospermae clearly exhibit an advanced level of siphonogamy. The development of a stigmatic area in the Magnoliales, as reconstructed by BAILEY and SWAMY (1951), may be taken as an indication of a different semophyletic sequence in this group. In *Degeneria* and *Drimys*, the pollen is caught on the matted hairs which close the slit of the carpel and the pollen tubes penetrate through the meshes between the hairs into the ovarial chamber. One could visualise a semophyletic development of these follicular gynoecia in which, originally, each individual ovule caught the pollen on one of its protective covers, and that gradually the pollen tubes became longer as the access of the pollen grains to the ovules became more and more obstructed by the infolding of the

pseudo-sporophyll. The final closure of the carpel, the subsequent formation of a stigmatic crest and ultimately of a style, were apparently not induced by the presence of a discrete placental stylar extension, but the
individual ovules may have had an apical tubular protrusion (tubillus).
The elongate chlamys (aril) of some Annonaceae, usually taken for the
outer integument (*e.g.*, by CORNER 1949) is suggestive in this connection.
After the gradual closure of the carpel prevented the direct pollination of
the ovules, their apical extensions became obsolete and often more or less
completely reduced.

All these considerations unequivocally point to a polyrheithric development of the carpellate gynoecia from aggregates of bracts and gynoclads
which became united into various 'apocarpous' or 'syncarpous' structures
with different forms of placentation. This is compatible only with a development of closed carpellate gynoecia along a broad front as the result
of a repeated homoplastic evolution. The semophyletic relationships between various categories of carpellate ovaries are often indirect in that
a common prototype was still ecarpellate, which implies that phenetically
similar types may have originated by means of a parallel development in
independent evolutionary lines. Considering that secondary changes,
adnations, concrescences, torsions, secondary syncarpy, peri- and
hypogyny, various reductions, etc. complicated the gynoecial morphology
still further, the interpretative and comparative analysis of the angiospermous pistils seems to become extremely difficult, but this is not a
serious disadvantage. The relationships of angiospermous taxa must be
established on the ground of the combined data from as many sources of
evidence as possible, and the gynoecial morphology apparently is not of
such paramount importance as its diagnostic value in applied systematics
suggests. Phylogenetic, anatomical, histogenetic, palynological, phytochemical and other indications (such as host-parasite relationships: cf.
LEPPIK 1957) gain in significance by being no longer obscured by conventional rules or dogmatic 'impossibles'. The typological derivations of
the various categories of gynoecia were always one-way streets, the
supposed sequence of semophyletic processes being thought to have
proceeded in one definite direction (*e.g.*, from apocarpy to syncarpy,
from marginal placentation to parietal and free central placentae), which
implied that the phylogeny of the taxonomic groups exhibiting the various
morphological levels of such a sequence coincided with the semophylesis
of the pistil and could only be read in that same direction. As soon as
these classical dicta are discarded, phylogenetic relationships become in
fact easier to derive because of the greater freedom of choice among
alternative possibilities.

The picture that emerges from this concise survey of the gynoecial
morphology is that of an evolution of the state of angiospermy along a

broad front and by means of various semophyletic processes. Sometimes the cupulate ovules themselves became the functional pistils, sometimes whole whorls of gynoclads and other bracts; sometimes a phalanx of ovules formed a common stigma or even a style, sometimes a gynoclad merged so completely with its bract that a new kind of organ, a pseudo-sporophyll, developed. The frequent association of the pistils with stamens and sterile perianth lobes in varying spatial relationships produced a multitude of functional 'flowers'. The advent of the closed carpellate gynoecium must have had such an impact on the evolution of the Angio-sperms as to have accelerated the evolutionary processes to the degree that intermediate stages in the semophylesis of the closed carpellate pistil have hardly been preserved among the recent Flowering Plants.

17

Phylogeny of the Microsporangium and the Stamen

Unlike the megasporangium, the microsporangium underwent only rather insignificant changes during the phylogeny of the Cormophyta and is, in all higher groups, still a closed saccate organ that produces microspores (at a higher evolutionary level, generally called pollen grains) which are subsequently released. Characteristic is the aggregation of the microsporangia into 'sori' or synangia, which is associated with fusions of the primary sporangiophores and frequently involves axes of a lower order on which the sporangiophores are inserted. Such groups of microsporangia, with their supporting axes, often became discrete morphological units of a higher order which in their turn evolved along various semophyletic pathways. The semophyletic changes predominantly involve these axes and not so much the terminal sporangia or their synangial aggregates.

As the prototype of the stamens of the Angiosperms, WILSON (1937) postulated a dichotomously branched telome system with terminal sporangia; LAM (1948 et seq.) accepts this basic type for 'stachyosporous' forms, and branched axial structures with terminal sporangia are also the basis of MELVILLE's (1960) gonophyll theory. However, this retrograde deduction, based on the general principles of the telome theory, relates advanced androecial structures of the most derived groups of the Cormophyta to very primitive 'psilophytic' conditions and, although the assumed connection is conceptually in order, the enormous 'jump' does not contribute any useful leads concerning the long and, as we shall see, varied evolutionary history. Primitive aggregates of sporangia resembling the postulated telomic (psilophytic) prototype occurred in such Devonian forms as *Svalbardia* and *Protopteridium*, it is true, but I believe that the Progymnospermopsida will serve much better as the starting point of the semophyleses of the microsporangiate reproductive organs. The assumption that *Svalbardia, Protopteridium* or taxa with a similar organisation may be precursors of the Progymnosperms is irrelevant in this connection, because only the morphology of the latter is fundamental in the step-by-step construction of the evolutionary sequences leading to the higher gymnospermous (and angiospermous) groups. The most important feature of the sporangiate structures in the Progymnosperms is that they are already more or less completely overtopped and appear as biaxial units, in *Archaeopteris* with the individual sporangia and sporangiophores free or mostly free, in Aneurophytales showing a marked tendency towards aggregation of the sporangiophores and sporangia into sori or synangia. These units were borne on a common axis, the rhachis of a mixed (partly fertile, partly sterile) pinna of the complex compound frond prevalent among the Progymnosperms, which also supported sterile assimilatory pinnules (see Figs. 1, 5 and 6). The phylogeny of the male reproductive organs of the Coniferophytina will be discussed in another chapter. The condition in the Aneurophytales provides a logical starting point for the semophylesis of the microsporangiate organs of the Cycado-filices proper, of the Euramerican seed ferns, which essentially retained the morphology of the Progymnosperms, in that they possessed compound mixed fronds. The tendency to form synangial complexes is very pronounced; the process was attended by a cladodic development of the axis supporting the usually sessile or subsessile synangia, which in some Lyginopteridales thus assumed the shape of a hairbrush, the sporangia being inserted perpendicularly to this axis (as in *Crossotheca*). In the Neuropteridales (Whittleseyinae), the morphology of the synangia be-came still more complicated, clavate, bell-shaped, semi-globose or dif-ferently shaped aggregates of the mostly intimately concrescent sporangia having been recorded. In the seed ferns of glossopteridalean affinity, ultimately, a similar soral or synangial association and a cladodic modifi-cation of some supporting axis occurred, which in their cycadopsid descendants often have the same basic morphology still, but there is no reason to conclude that these cladodic (laminose) sporangiate organs are morphologically quite comparable to the male reproductive organs of the Euramerican Cycadofilices, because in such groups as the Cycadales the laminose organ bears numerous small groups of sporangia, whereas the peculiar male reproductive organs of the Cycadeoideales (which are *not* pinnate in *Cycadeoidea* as suggested by the figures and descriptions in numerous textbooks and manuals, all based on reconstructions made by WIELAND: see DELEVORYAS 1963) clearly possessed compact synangia in a range of few to many. The question arises as to whether these organs or only the individual synangia are homologous with a microsporangiate organ of the *Crossotheca* type; in other words, are these laminose spo-rangiate organs of the protocycadopsid and cycadopsid groups biaxial structures or are they triaxial? The answer is most probably that there is no essential difference because the association of all sporangia based on the same axis into a single soral or synangial complex or into a number of such complexes are likely alternatives. A fully comparable situation is known in the female reproductive organs of the pteridospermous and protocycadopsid groups: the cladodic axis supporting the synangia (ovules), the cupule, sometimes contains a single ovule and in other cases two or more. It seems best to base all homologies and semophylcscs on a biaxial sporangiate structure, but such units occurred in groups on a common main axis and form triaxial aggregates (as in *Crossotheca*).

The frequent occurrence of laminose male organs among cycadopsid groups has in the past strengthened the postulated sporophyll concept, because the phenetic similarity between the supposedly primitive laminose stamens of the Magnoliales and the microsporangiate organs of other spermatophytic groups (including the unfortunately misinterpreted 'microsporophylls' of *Cycadeoidea!*) was suggestive of a phylogenetic and semophyletic relation. By far the large majority of the Angiosperms (and all the Chlamydosperms) exhibit the axial nature of the stamens in the slender filaments topped by the synangium (the anther), and I believe that in the 'Polycarpicae' with 'leaf-like' stamens, such as Magnoliales and Nymphaeales, these laminose microsporangiate organs are pseudo-phyllosporous organs of dual origin. The broad and seemingly foliaceous atypical stamens of Cannaceae, Zingiberaceae and Iridaceae are not only manifestly homologous with the stamens of the Liliiflorae but also antitepalous, so that I believe that they are cases of secondary cladodification of the general filamented type of stamen.

The difficulty of the derivation of the typical angiospermous stamen from a laminose (cladodic) protocycadopsid or cycadopsid archetype is only apparent, because there are some Mesozoic groups with non-laminose microsporangiate organs, *viz.*, the Caytoniaceae and the Pentoxylales, and one might attempt to explain this alternative condition as the retention of the most primitive (protogymnospermous) form of androecium in some cycadopsid taxa. However, purely morphological considerations may obscure the recognition of the selective pressure of the adaptation to entomophily which, according to several authorities, became firmly established in the Mesozoic period. The polliniferous organs (the sporangia or synangia) must have been, or have become, sufficiently exposed to be accessible to the more ancient pollinating insects such as beetles and it is conceivable that laminose organs became secondarily reduced to slender axial structures and the synangia became long-stalked as well as fewer in number. If this is accepted, the more or less direct phylogenetic derivation of the angiospermous stamen from a dichotomously branched telome system (WILSON, MELVILLE) appears to fall far short of the more likely sequence of (1) dichotomously branched psilophytic sporangiate telome systems→(2) biaxial units ('fertile pinnules') borne on a common overtopping axis (Progymnosperms)→(3) the laminose cladodification of the axis of the biaxial unit (pinnule) and aggregation of the microsporangia into synangia (Pteridosperms, Protocycadopsida, Cycadopsida, partly)→(4) the secondary reduction of the laminose organs (cf. *Caytonia*)→(5) a reduction and oligomerisation to secondarily stalked microsporangiate organs or stamens (Pentoxylales, Chlamydosperms, Angiosperms; see Fig. 15).

A constant feature is the association of the reproductive organs with a subtending foliar organ or bract. I agree with MELVILLE that (in cycadopsid groups at least) the fertile organ and its associated bract constitute a unit reflecting an ancient connection, but such assumptions can be substantiated only by tangible semophyletic archetypes, so that the common origin of a gonoclad-bract unit must be sought in a mixed pinna of the progymnospermous compound frond (MEEUSE 1963a) rather than in hypothetical forms with dichotomously branched 'primitive gonophylls' (MELVILLE). The presence of sterile assimilatory organs and fundamentally biaxial fertile organs on the same axis (see Fig. 4) strongly suggests that the gonoclad and its bract are derived from such a progymnospermous sporangiate structure by a shift of the position of the sterile elements, so that an axial relationship developed. The gonoclad was apparently primarily triaxial, its ramifications consisting of (1) the original central axis (rhachis) of the mixed pinna, (2) secondary axes (each representing a secondary rhachis of a 'fertile pinnule'), and (3) the sporangiophores. Such an arrangement of the androclad can be recognised in several forms (*e.g.*, in Cycadales, *Pteruchus* and *Caytonia*), but later in the phylogeny of the Angiosperms there must have been a progressive oligomerisation and reduction of the secondary and tertiary axes which became remoulded into a single *stalked* synangium or stamen, so that ultimately the androclad became seemingly biaxial. Analogous reductions of the gynoclads transformed a triaxial system into an ovuliferous axis bearing stalked arillate ovules which is also seemingly biaxial and, if the comparison is permissible, many traditional stamens may well be condensed biaxial structures, but their present morphology and anatomy usually no longer betray their origin, unless the dichotomously branched 'stamens' of, *e.g.*, *Ricinus* are supposed still to reflect the complex nature.

Despite several uncertainties in this sketch of its semophyletic history, a fundamental microsporangiate unit emerges which is the bracteated triaxial or, for practical purposes, ultimately biaxial androclad (see Fig. 16). This complex morphological structure is the basic androecial element of the Angiosperms, which in many lines subsequently underwent further oligomerisations and reductions culminating in the solitary stamen (frequently still subtended by a bract!). As a rule, several androclads and the corresponding bracts are inserted on a common axis, with or without a number of gynoclads, to form a reproductive unit of a higher order, the anthocorm. This interpretation of the microsporangiate organ and higher compound structures provides the basis for a new floral morphology, but there are several complications which will be discussed in a special chapter. A slight semantic difficulty is caused by the traditional term 'stamen' which is, in the new context, a stalked synangium, which

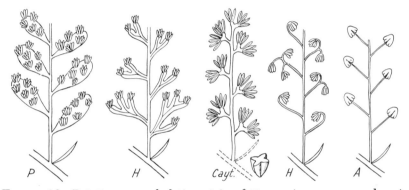

FIGURE 16. Putative semophyletic origin of the angiospermous androecium from a protocycadopsid archetype: *P*—Protocycadopsid androclad consisting of flat organs (male counterparts of ovuliferous cupules) supporting a number of male sori of microsporangia (or androsynangia) on adaxial (or abaxial) side; compare Cycadales, *Pteruchus*. *H*—A hypothetical phase indicating reduction of organ supporting the androsynangia (prototypes of thecae of stamens), leading to the condition in Caytoniales (*Cayt.*), another hypothetical chlamydospermoid-protangiospermous type, and primitive angiospermous androclad bearing numerous stamens. In several groups an extreme oligomerisation reduces the stamens to a single one per androclad. More primitive (multistaminate) androclads are usually recognisable because they frequently develop centrifugally (compare Fig. 17).

means that a number of conventional 'stamens' belonging to the same androclad are not the absolute morphological equivalents of a 'stamen' which represents a reduced (extremely oligomerised) *whole* androclad. In order to avoid any confusion when the familiar term is to be retained, one must simply define the stamens as the derivatives of primary synangia (the ultimate ramifications of the androclad) so that one can also describe the basic type or form of an androclad of the Angiosperms as a (seemingly) biaxial structure of which the branches are formed by stamens, oligomerisation of the number of stamens eventually leading to the single bracteated (antitepalous, epipetalous or antisepalous) stamen. In practical interpretative floral morphology, the relation of the androclad (*i.e.,* either a single stamen or an assembly of stamens) to the subtending bract is the best guiding principle, because this bract has only seldom disappeared. The traditional typological interpretation of the multistaminate androecia of such forms as Dilleniales, Guttiferae, Cistales, Myrtales, Malvales, etc., was based on the assumption that the stamens are normally arranged in alternating whorls, so that 'irregularities' in the supposed series of whorls had to be explained by the reduction of a whorl (in cases of 'obdiplostemony') or by serial splitting (*dédoublement, chorisis*). 'Splitting' has always been a moot point in the classical floral morphology and was actually nothing but an ancillary postulate to save the theory. The presence of associated stamens (in the pentamerous flowers of the dicotyledonous groups under discussion, usually forming five or sometimes ten groups) merely reflects the retention of the primitive biaxial androclad with numerous individual androsynangia (stamens) and clearly demonstrates that these dialypetalous taxa cannot possibly have descended from the Polycarpicae, which almost invariably have androclads reduced to single (bithecate) stamens, and have, accordingly, more advanced androecia than the former. This is one of several indications (including the pollen morphology!) of the independent origin of such groups as Magnoliales, and of the early specialisation which tended to make these forms morphologically rather rigid and at the same time conservative. The order Magnoliales comprises the 'terminal' members of an old taxon which shows signs of being well on the way to extinction. The androecial morphology of the Monochlamydeae is very complicated because early progressive reductions and oligomerisations vie with tendencies to retain ancient structures. The comparative floral morphology of the recent Amentiflorae has thus become so difficult that the mutual relationships within this group are obscured.

A progressive oligomerisation of the androclads is a regular phenomenon among nearly all advanced angiospermous groups which is most probably associated with the increased efficiency of the pollination process in zoophilous flowers, the greater precision requiring far smaller quantities of pollen than in anemophilous forms. The androclads are usually reduced to a single stamen, and the number of androclads per anthocorm ('flower') has often gradually decreased; for example: at first to six, subsequently to three stamens in numerous monocotyledonous orders (Iridales, Cyperales, Poales, etc.), and eventually to two or a single one in Orchidaceae and some Scitamineae; to four or five (or eight to ten, rarely three or six) in many dicotyledonous groups (Caryophyllaceae, Cornales-Umbelliflorae, some Rosiflorae, Geraniales, Berberidaceae, Menispermaceae, and practically all sympetalous orders) culminating in such families as Oleaceae (flowers mostly with two stamens) and

Valerianaceae (three, or only one stamen) and in some Anacardiaceae, Vochysiaceae, Labiatae, Scrophulariaceae, Acanthaceae, etc. (with two stamens or occasionally only a single one). The number of perianth lobes usually betrays the original number of androclads insofar as the sepals, petals or tepals represent the bracts of the androclads.

One characteristic and fairly constant feature of the stamen is the bithecate (tetrasporangiate) anther. This is presumably an ancient feature—it occurs in both Monocots and Dicots!—and the often tetra-sporangiate synangia in Caytoniaceae may be phylogenetically significant in this connection. Different types of anthers among the Angiosperms are of special interest, because they may reflect alternative ancient condi-tions such as bisporangiate synangia or solitary stalked microsporangia ('fertile telomes'), but the possibility of a secondary reduction or segrega-tion must not be precluded. The stamens of the genus *Salvia, e.g.*, clearly demonstrate that a proliferation of the connective may cause the spatial separation of the thecae, and the more or less complete super-position of the thecae in many Acanthaceae-Justicieae also indicates a certain (perhaps secondarily acquired) independence of the elements of the synangium. A reduction of one of the thecae in such cases would produce a monothecate anther. The anthers of the Malvaceae are not typical of the Columniferae (Malvales) and their phylogenetic sig-nificance appears rather small; similarly, the 'bifurcate' stamens of the Betulaceae are exceptional among the Amentiflorae and transitions to undivided stamens occur in the same family (in *Alnus*), so that a second-ary segregation (splitting) of the thecae is quite feasible. This kind of splitting is a real division (a bisection) and of course quite different from the postulated morphological 'splitting', *i.e.*, replication of complete ele-ments, mentioned previously in this chapter.

The peculiar branched androecial elements of *Ricinus* and a few other Euphorbiaceae have terminal ramifications topped by bithecate anthers, so that, although these branched structures simulate a dichotomously bifurcating telome system, their terminal portions (the synangiophores) represent at least four fertile telomes and the dichotomy is not necessarily a remnant of a primitive mode of branching. A dichotomy or pseudo-dichotomy is also present in the vascular supply to the multistaminate androclads in Dilleniaceae, *Paeonia*, Guttiferae, Myrtales, etc. (see, *e.g.*, SPORNE 1958), but vascular traces often bifurcate and the branching pattern of such forms could also be called 'dendroid', sympodial or mono-podial. At any rate, the combined evidence is rather indicative of the synangial status of the primitive anther than of a dichotomously branched system of bi- or monosporangiate axes, so that I do not think the inter-pretation of the primitive stamen as a bifurcating system of telomes (WIL-

SON 1937) is of much practical importance. Another pitfall is the false homology of all 'stamens', *i.e.*, the supposed equivalence of a single, terminal bithecate ramification of a branched androclad and a stamen which represents a whole oligomerised and reduced androclad. One can compare only the androecia at the semophyletic level of primitive branched (*i.e.*, 'multistaminate') androclad, because the phylogeny of the androecia of the cycadopsid groups really begins with strobiloid triaxial associations of synangial structures as we have seen, the next step being the advent of the seemingly biaxial multistaminate androclad.

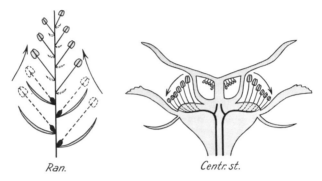

Ran.   *Centr. st.*

FIGURE 17. Comparison of centripetal and centrifugal development of stamens, at the same time demonstrating the considerable fundamental differences in morphology between a ranunculaceous form (*Ran.* = *Ranunculus*) and a dilleniaceous-paeoniaceous type of flower with centrifugal stamens (= *Centr. st.*, based on a median longitudinal section of a flower of a dilleniaceous type of plant), in spite of the superficial phenetic resemblance. In *Ranunculus* (*Ran.*) the functional stamens each represent an androclad reduced to a single stamen, the subtending bracts being reduced (the presence of a kind of leaf-gap structure in the stelar anatomy of the floral axis opposite the base of the filament supports this interpretation), so that the stamens develop acropetally (*see arrows*). In the Dilleniaceae and in *Paeonia* the stamens occur in groups, each group being based on a common main vascular strand opposite a petal and (manifestly) a multistaminate androclad, the petal representing the supporting bract. The androclad being a 'subsidiary' axis of main floral axis, the morphologically proximal portion of the androclad is near the centre of the flower and the development would normally proceed acropetally along the subsidiary axis in the direction indicated by the arrows, i.e., 'centrifugally'.

The 'centrifugal' stamens (CORNER 1946) occur in families with such multistaminate androclads: Paeoniaceae, Dilleniaceae, Bixaceae, Guttiferae, Myrtaceae, and several others. The explanation of this phenomenon is manifestly that the most proximal stamen of an androclad, which is closest to the centre of the flower, represents the lowermost lateral branch of the ramified axial system and matures first if (as is likely), the differentiation of the individual stamens proceeds acropetally along the main

androclad axis, the visible effect being a centrifugal sequence of staminal development (Fig. 17, *left*). Significantly, in flowers with mono-
staminate androclads (Ranunculaceae!) the development is centripetal,
the lowermost androclads (stamens) being the first to be formed on the
floral axis and the development progressing acropetally (Fig. 17, *right*).

18

Problems of the
Polyaxial Floral Region

The general picture emerging from the considerations and deductions concerning the morphological nature of the reproductive structures of the cycadopsid Cormophyta is that a number of sporangiate organs, the gonoclads, are inserted on a common axis and thus form a unit of a higher order, the anthocorm. Several elementary questions concerning this basic structural unit must be satisfactorily answered before it can serve as the common denominator in a new comparative and interpretative floral morphology.

As regards the basic homology and the phylogenetic origin of the floral region, if one accepts the view that the gonoclads and their bracts are derivatives of a 'mixed pinna' of a progymnospermous 'compound frond', the assembly of a number of such pinnae on a common rhachis is a primary condition which means that a whole compound frond, or at least a large portion of it comprising several of its sporangiate pinnae, became transformed into a polyaxial reproductive structure, the anthocorm. The gonoclads, after a more or less varied evolutionary history, evolved into the androecia and the gynoecia of the Angiosperms, so that a continuous semophylesis of the reproductive organs from the Upper Devonian to recent time can be postulated. Another advantage of this point of view is the possibility of aligning the comparative morphology of all derivatives of the Progymnosperms, *i.e.*, of all gymnospermous and angiospermous groups, by basing it on a common archetype and by using a terminology applicable to all forms. The conventional and biased terms 'strobilus', 'cone', 'flower', 'inflorescence', 'sporophyll', etc., must be replaced by the neutral and phylogenetically (semophyletically) sound terms anthocorm and gonoclad.

The second point of inquiry concerns the relation of the male and the female sporangiate organs, the old controversy of the unisexual or bisexual primitive 'flower'. It is known that the conditions of monocliny and dicliny are not only genetically (*i.e.*, phylogenetically) determined, but also subject to physiological, pathological and other external influences which may change the situation (see, e.g., HESLOP-HARRISON 1958). From an evolutionary point of view, these physiologically controlled alternations of the sexual condition in a reproductive region can be important only if they are to some extent hereditary, so that in a morphological discussion such induced changes of sex had better be disregarded (which does not mean that they are not significant in many respects!). A superficial survey of the Gymnosperms, both living and fossil, reveals the predominance of unisexual reproductive regions (dicliny), but the exceptions are all found in the over-all line of descent leading to the cycadopsid forms: Glossopteridales (according to PLUMSTEAD, 1952, 1956a, 1956b), Marsileales, Cycadeoidales, and (morphologically) Chlamydospermae. The situation is somewhat more complicated, however, because one must distinguish between the occurrence of unisexual (homosporangiate!) anthocorms and bisexual (amphisporangiate) anthocorms, and between the occurrence of *compound* anthocorms comprising unisexual anthocorms of the one sex only and of compound structures with unisexual anthocorms of both sexes. After considerable reductions and oligomerisations, the compound anthocorm bearing both male and female elementary anthocorms may simulate a bisexual structure (cf. *Cryptocoryne*). Finally, in the Angiosperms at least, the occasional occurrence of dicliny in normally monoclinous taxa (*e.g.*, Labiatae, Compositae, Orchidaceae) indicates the likelihood of a secondary monosexual condition derived from a bisexual one by the suppression of the genitalia of the one sex. The more primitive angiospermous groups do not exhibit a consistent pattern. Magnoliales, Piperales, Ranunculales, Dilleniales and Liliiflorae are mostly monoclinous; Monochlamydeae, Pandanales and Arecales almost exclusively diclinous; and some orders are more or less 'mixed' (Arales, Cyperaceae, Rosiflorae, Guttiferae, Parietales). In view of the probable multiple origin of the Angiosperms, the inevitable conclusion is that the various forms of distribution of the sexes are essentially *alternatives* already co-existing at the pre-angiospermous level of evolution. The increase in the number of monoclinous forms must have been the result of the positive selective effect of the bisexual condition when insect pollination became more general (a single 'visit' to a bisexual flower is usually sufficient to achieve both the transfer of pollen to the insect and the reception of extraneous specific pollen from the same carrier). Significantly, anemophilous Angiosperms, of which the (most probably) secondarily re-adapted forms such as some Ranunculaceae (*Thalictrum*) and Compositae (*Artemisia*) provide the most interesting examples, are frequently diclinous (Monochlamydeae, Pandanales, Arecales, Cyperales). The phytomorphologist need not and must not make a major issue of the question whether, in the Angiosperms, the monosexual or the bisexual reproductive region ('flower') is the more 'primitive'. The advent of the bisexual primitive anthocorm preceded the development of the first full-fledged Flowering Plants and the evolution of the conventional 'flower' (compare also the discussion in HESLOP-HARRISON 1958).

A practical task to be accomplished is the 'translation' of the present-day morphology of the reproductive regions into the terms of the general basic pattern, not only for purely morphological and phylogenetic reasons, but also to achieve the replacement of the standard typological methods in taxonomic classification by a more realistic approach. Since it is not to be expected that all intermediate links will ever be found as recognisable fossils, and also for practical reasons, a new typology of the floral region is indispensable if systematics, morphology and phylogeny are to be kept in line as much as possible. Provided general phylogenetic (semophyletic) tendencies can be recognised, typological derivations of the floral regions from anthocorms, based on these trends, are a much better approximation of the actual evolutionary processes than the conventional deductions starting from a monaxial (and usually bisexual) 'flower' with foliar fertile appendages, and at least indicate unequivocally the sequence of the morphological changes (i.e., the semophyletic series can be read in one direction only). A recognition of these trends is of course equivalent to the solution of the question of how to relate all recent floral structures to the postulated prototype.

The first assumption to be made is that there are three possible alternative conditions, viz., (1) a functional reproductive entity (in traditional terms: a 'flower') may be the derivative (homologue) of an anthocorm, (2) it may be the equivalent of a fertile element of an anthocorm, and (3) it may represent a number of anthocorms inserted on a common axis (a compound anthocorm). The first point to be decided is whether all the functional 'flowers' represent only one of the three cases or if they are the respective homologues of two or three different ancestral structures. The morphological identity of a 'flower' with a whole anthocorm can easily be ascertained by a point-for-point comparison of the two. An anthocorm consists of a common central axis bearing a number of gonoclads supported by bracts and usually provided with a number of sterile phyllomes proximally of the sporangiate organs. A gynoclad and the associated bract are often transformed into a carpel; an androclad appears as a group of stamens or a single stamen generally opposite a perianth lobe (usually a petal); and the basal phyllomes are represented by some or all of the perianth lobes, or by the perigone, the calyx or the prophylls. If the various elements are aggregated into a discrete higher unit with a common supporting axis, the homology of this functional 'flower' with an anthocorm is manifest. The conventional flowers of the Nymphaeales, Magnoliales, Ranunculales, Centrospermae,

Parietales, Guttiferae, Juncales, Scitamineae, Commelinales, Helobiae and their derivatives (e.g., Sympetalae, Orchidales), consisting of a 'carpellate' gynoecium, an androecium and a perianth (or perigone), all belong to this group, and the absence of one of the categories of fertile elements (carpels or stamens) does not make any difference in this connection, so that the unisexual flowers of, e.g., Euphorbiaceae, Cucurbitaceae, Begoniaceae, and Araceae also belong here, the interpretation being that the unisexual flowers are either the homologues of primarily unisexual anthocorms (as in Araceae and Begoniaceae), or the derivatives of originally bisexual anthocorms in which the one sex is reduced or suppressed (as, perhaps, in Lardizabalaceae and Menispermaceae). The 'minimum requirements' for a 'flower' to qualify as a modified anthocorm are (1) an aggregate of carpels (a solitary carpel is suspect, because it represents only a single gynoclad and not necessarily a remnant of an assembly of carpels that underwent a secondary oligomerisation) or (2) a whorl of stamens with at least an isomerous simple perigone (an aggregate of 'naked' stamens, or a solitary stamen without a whorled perianth may be homologous with a single androclad), not necessarily both (1) and (2) because an anthocorm may bear gonoclads of only one sex (and the associated bracts).

As we have seen, not all angiospermous taxa have carpellate gynoecia, and retained the primitive gynoclad (the cupuliferous axis), almost invariably with its subtending bract. If such a gynoclad became reduced to a solitary bracteated and one-ovuled pistil, as in Cyperaceae, the gynoecium of the reproductive structure derived from the anthocorm consists of a few of these pistils in a whorl primarily surrounded by bracts (some palms and an occasional Cyperacea); such pistils have sometimes coalesced (Arecaceae generally, Restionales, secondarily again reduced in Poaceae to a mono-ovulate 'pistil'). In bisexual forms, these oligomerised and reduced gynoecia are surrounded by stamens and their bracts (perianth or perigone, see Scirpoideae, grasses and the monoclinous palms). The origin of such floral structures from a whole male, female or bisexual anthocorm permits the qualification 'flower' for such reproductive units, if one redefines a 'flower' as the semophyletic derivative of an anthocorm, which is the obvious thing to do, because in this way the large majority of the Angiosperms can be said to have flowers and the status quo can be maintained as much as possible. Even the undoubtedly much reduced functional flowers of the Monochlamydeae need not form an exception, if one assumes that the androecia and the pistil are the remnants of whorls of gynoclads. However, there are several taxa, which have reproductive units traditionally called 'flowers', that are not readily brought under the new definition of the flower as a homologue of an anthocorm. Especially the acyclic staminate 'flowers' of several Amentiflorae are not necessarily the derivatives of helically constructed anthocorms. The floral region of the Amentiflorae has been the subject of an impressive number of studies (*e.g.*, by ABBE, ABBE and EARLE, FISHER, MANNING, LANGDON, LEROY and especially HJELMQVIST 1948), but the interpretative morphology of these reproductive structures was approached in the traditional manner, and it is irrelevant if certain functional entities were treated as reduced euanthia or as pseudanthia, the primary supposition in either case being a reduction of a more complex structure (a primitive 'flower') to, say, a bract and one or more 'naked' genitalia. MELVILLE (1960) has suggested that the so-called flowers of some Amentiflorae are single 'gonophylls', *i.e.*, structures resembling gonoclads subtended by bracts, and it is indeed tempting to compare the simple amentiflorous staminate structures with a primitive androclad (represented by a group of sometimes more or less coalesced stamens) and its bract. (Incidentally, MELVILLE sees an archaic feature in the frequent insertion of the filaments on the bracts, supposed to reflect a postulated primitive 'epiphylly' of the genitalia). However, most authorities are agreed upon the extreme reductions in the fertile regions of the Monochlamydeae and I concur in so far, that the oligomerisations, condensations and other simplifications must have set in when these regions were still in the primitive prefloral phase. Accordingly, phenetic resemblances must be regarded with considerable caution, because there is *a priori* no reason to preclude the extreme reduction of a whole anthocorm to a single gonoclad (without its bract!), only the bract of the *anthocorm* being retained, and thus simulating the presence of a bract subtending the 'naked' gonoclad (a rather similar case is known in the Caricoideae, as we shall see). Corroborative evidence is obviously needed, and the comparative morphology of the Amentiflorae comes first to mind as a promising line of inquiry. Although the comparison of different taxa seems to be the logical method of approach, the unequal degree of reduction to be anticipated renders the selection of the most primitive conditions too hazardous. A much more promising method of analysis is the point-for-point comparison of the male and the female regions, the fundamental correspondence of micro- and megasporangiate anthocorms being germane to the recognition of the various categories of bracts and axes.

If the Salicaceae are taken as the first example and their gynoecia are interpreted as carpellate structures, the pistils provide the first clue, because there are, accordingly, two or more placental axes representing a whorl of gynoclads. The axis on which the pistil is inserted is the axis of the female anthocorm and the short stipe ('pedicel') of the ovary would qualify if the nectarial glands (*Salix*) or the entire two-lobed 'disc' (*Populus*) inserted on this same stalk did not complicate the interpretation. There are three alternatives (at least, if the interpretation of the nectaries as glandular enations without morphological value is disregarded), *viz.*, the nectarial and disc-like organs represent (1) reduced pistils, (2) vestigial gynoclads or gynoclad bracts, or (3) transformed (proximal) sterile phyllomes of the same anthocorm on which the functional pistil is coaxially inserted. If they are rudimentary pistils, the traditional 'female flower' would be a very complex structure comprising several anthocorms (as many as there are pistils and nectaries, or pistils and disc lobes, respectively) based on a common axis, in other words, a compound anthocorm. In the other two cases the presence of only a single anthocorm need be assumed. Abnormalities ('teratologies') including those observed in the occasionally formed summer catkins suggest a foliar nature of the nectarial elements (for a detailed discussion, see HJELMQVIST 1948, pp. 150–152) and this would rule out their homology with a whole pistil (anthocorm). It is irrelevant to the present problem whether the accessory vestigial elements represent sterile basal bracts or gonoclad bracts, because in either case they indicate that the pistillate 'flower' is a condensed and reduced female anthocorm. The female ament is, according to this interpretation, a compound anthocorm and its male counterpart, which is borne in a corresponding position, is also fundamentally a compound anthocorm, each staminate 'flower' representing a whole anthocorm. It is only of academic interest whether the androecium is derived from a single remaining multistaminate androclad or consists of a number of androclads, each reduced to a single (or a few) stamens (there are no developed supporting bracts to guide us), *i.e.*, whether the number of androclads or only the number of stamens per androclad underwent an extreme oligomerisation. However, the occasional occurrence of nectarial glands—which, as we have seen, may be the transformed bracts of (reduced) androclads—*amidst* the stamens in *Salix* pleads in favor of the second assumption. The androgynous flowers occasionally encountered suggest the equivalence of parts of the gynoecia and the stamens, and this could be interpreted as the equivalence of a stamen with a gynoclad, *i.e.*, the stamen represents an androclad, which conclusion also supports the second alternative.

A consequent morphological analysis based on the assumption that the salicaceous gynoecium is carpellate thus proves the deceptively simple catkins of the Salicaceae to be complex structures, derived from compound anthocorms, and, if the new concept of the flower as a modified anthocorm is somewhat stretched, each functional reproductive unit can be called a 'flower'. In my opinion there are, however, insufficient grounds to accept this interpretation as final. If the salicaceous pistil is ecarpellate and represents a pluri-ovulate cupule homologue, the so-called female flower is a condensed gynoclad and the catkin an anthocorm. In the same way, assuming that the Fagaceae have carpellate gynoecia, the identity of each placental axis with a gynoclad provides the clue to the interpretation of a 'female flower' as a modified unisexual anthocorm and the aggregates of pistillate flowers as compound anthocorms which are sometimes encased by the 'cupule', an (axial) derivative of the common peduncle. The 'male flowers' are usually 'cyclic' and in this interpretation clearly represent a whorl of androclads and their subtending bracts. The conditions in the Betulales are very much the same. A detailed discussion of all amentiflorous orders remains premature as long as there are two possible interpretations of their gynoecia (and any reader so inclined can in any case work the floral morphology out along either of the alternative lines), but some remarks must be made regarding the Myricaceae and Juglandaceae. The gynoecia in these two families are essentially 'gymnospermous' bitegmic ovules and the most probable interpretation (MEEUSE and HOUTHUESEN 1964) is that the so-called 'female flower' is an arillate ovule (the 'calyx' or 'perianth' of the Juglandaceae representing the cupule or chlamys). It would seem as if a female catkin represents a gynoclad but this is manifestly not the case, because there is a supporting bract to each ovulate structure indicating that there is at least a secondary axis, i.e., a gynoclad reduced to a single subsessile ovule, in other words, the presence of bracts or modified (adnate or enveloping) bracts associated with the pistillate reproductive structures suggests the interpretation of these traditional 'flowers' as simplified (condensed) gynoclads. It will be clear that the new approach to the comparative floral morphology of the Amentiferae outlined in these paragraphs, even if alternative interpretations of the morphology of the pistil may complicate matters in some families, is much more fruitful than previous attempts and 'makes sense'. At the same time, MELVILLE's interpretation of the 'male flowers' of Salicaceae and Corylaceae as 'androphylls' (i.e., bracteated androclads) is thus proved to be debatable, to say the least. The extreme oligomerisations, reductions and condensations in the reproductive regions of the Monochlamydeae also indicate a special evolutionary trend which must have started in their primitive (perhaps still hemi-angiospermous) prototaxa at an early evolutionary stage. All attempts to derive the assembly from dicotyledonous groups with more or less 'complete' flowers are entirely futile—the Monochlamydeae constitute an ancient and independent line of descent with a progressive simplification of the anthocorms in the 'prefloral' stage.

Early simplifications of complex anthocorms must also have been the prevailing evolutionary trend in the Cyperaceae. If one assumes an oligomerisation of the androclads to single stamens, frequently attended with a loss of their bracts and a reduction of the gynoclads to solitary (chlamydote) ovules, the bisexual flower of the Scirpoideae can be derived from a considerably simplified bisexual anthocorm, and the floral structures of the Caricoideae from compound anthocorms (generally bearing unisexual elementary anthocorms), as indicated in the accompanying diagrams (see Fig. 18).

A typology of the floral regions of all Angiosperms based on a common prototype, the anthocorm, can be construed, if one postulates that certain general phylogenetic tendencies were operative and changed ele-

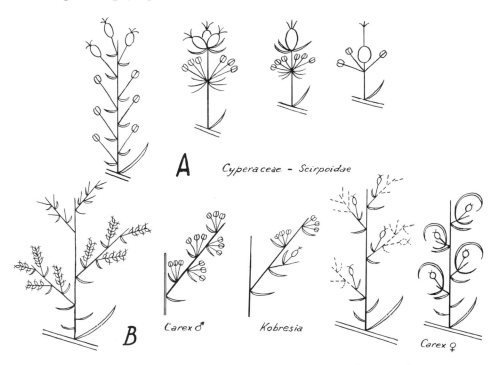

FIGURE 18. Tentative semophyletic development of the reproductive regions of the Cyperaceae: A—Scirpoideae. In the more primitive forms, the ambisexual anthocorm bore modified gonoclads reduced to a single one-ovuled pistil of cupular derivation in the female region and to a single stamen in the proximal male region (cf. *Scirpodendron*), followed by an oligomerisation of the gynoclads to a single one-ovuled pistil and of the androclads to a whorl of three (or two trimerous whorls), and by the reduction of the subtending bracts. B—Caricoideae. One must assume that the compound anthocorms in these groups bore male and female elementary anthocorms (progenitors of *Kobresia*, etc.) or only anthocorms of the one sex (progenitors of *Carex*, etc.). Each female anthocorm of the compound structure became reduced to a short bracteated axis bearing only a single bracteated gynoclad oligomerised to a single pistil; each male anthocorm became reduced to a triad of ebracteate sessile androclads, each oligomerised to a single stamen, the triad being supported by the anthocorm bract. In *Kobresia* and in other genera, the gynoclad bract is not closed into a utricle as it is in *Carex*.

ments of the anthocorm semophyletically in various and sometimes alternative directions, which need not always have occurred in the same order. The more important initial trends and specialisations which transformed an anthocorm into a conventional 'flower' or into a traditional 'inflorescence' are the following:

1. A tendency towards a cyclisation of the anthocorm, *i.e.*, the re-arrangement of the gonoclad-bract units and the sterile perianth lobes into 'whorls' or tiers.
2. A tendency towards oligomerisation of the number of the gonoclad-bract units of one or of both sexes, usually associated with a more or less absolute fixation (stabilisation) of the ultimate number of elements in each category of fertile and sterile organs.
3. A tendency towards oligomerisation of the number of male synangia (stamens) and/or of the number of cupules (chlamydote ovules) per gonoclad.
4. A tendency towards the formation of carpellate pistils and, much more rarely, of pseudo-phyllosporous androclad-bract units.
5. A tendency towards a reduction ('loss') of the fertile axial element of a gonoclad-bract unit or, conversely, of the sterile element (the bract).
6. A tendency towards a reduction of the main bracts subtending the primary anthocorms in compound anthocorms.

These trends, of course, did not develop in every taxon, because some of them are mutually exclusive or alternative semophyletic pathways; this provides several clues concerning the early evolution of various angiospermous groups and thus indicates a number of possible common, divergent, and alternative (parallel) phylogenetic lines that can serve as a yardstick of the degree of phylogenetic relationship. Some elucidating examples: If an oligomerisation of the androclads to a single stamen precedes other trends, the taxa with such simplified androecia cannot possibly be the progenitors of taxa which retained the more primitive multistaminate form of androclad. A possible relationship of the two groups of taxa can only be based on the assumption that their common progenitors had primitive anthocorms and that an incipient divergent trend in one line (the early oligomerisation) was responsible for their separation into (at least two) parallel lineages. The large dicotyledonous plexus of Dilleniales-Cistales-Theales-Guttiferales-Parietales-Myrtales-Rosiflorae-Malvales with primarily multistaminate androclads cannot possibly be *derived* from plants with the androecial morphology of Magnoliales or Ranunculaceae, but a common *origin* of Ranunculales and the more primitive members of the plexus from early progenitors with primitive anthocorms at the prefloral stage is not at all improbable. Similarly, an extreme oligomerisation of the gynoclads to solitary (chlamydote) ovules in one assembly (such as the Cyperaceae) not only precludes the derivation of forms with pluri-ovulate carpels (*e.g.*, Juncaceae) from the former, but also the semophyletic origin of the non-carpellate cyperaceous gynoecium from a carpellate juncaceous pistil. A possible relationship between such groups (say, between Cyperaceae and Juncaceae) can only rest upon their propinquity of descent from a common prototype with primitive (pluri-ovulate) gynoclads subtended by bracts, *i.e.*, at the precarpellary semophyletic level. The recent Pandanaceae (*Freycinetia* perhaps excepted) have lost the bracts subtending the gynoclads and they cannot have been the direct ancestors of the numerous monocotyledonous taxa with carpellate ovaries, nor of those which primarily retained the bracts although they did not attain the evolutionary level of the carpel (such as Arecales and Cyperales). The early Monocots presumably had

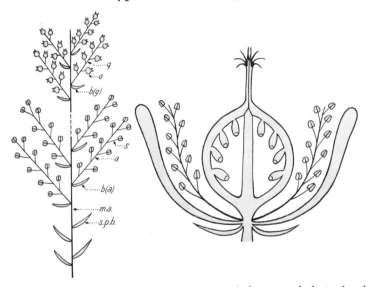

FIGURE 19A. Diagrammatic representation of the semophyletic development of an angiospermous flower from an anthocorm. The already somewhat advanced anthocorm (left) bears bracteate gynoclads (*g*) distally and androclads with numerous stamens (*a* and *s*), likewise subtended by a bract *b(a)*, more proximally, followed by some sterile appendages (*s.p.b.*). After oligomerisation of the number of gonoclads and subsequent cyclisation, a tetracyclic structure (see Fig. 15, *top*) has developed which becomes transformed into a 'flower' by the amalgamation of gynoclads and their bracts. In Fig. 15, the gynoecium is rendered as being syncarpous and as having 'axial' placentation (a condition found in Liliales, Guttiferae, etc.) and the androecium is shown as in Liliales, etc., with extreme oligomerisation of the number of stamens per androclad. In this figure (19A) the gynoecium is rendered as having 'parietal' or 'laminal' placentation, but the androecium is shown in the rather primitive condition found in, *e.g.*, Opuntiales, Dilleniales (and Paeoniaceae), Myrtales, Clusiales = Guttiferae; such androecia often develop centrifugally. Compare also Fig. 17, in which a similar diagram is shown which is reminiscent of the actual floral structure of the Dilleniales.

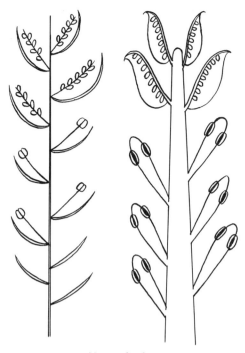

Magnoliales

FIGURE 19B. As in Fig. 19A, but the putative derivation of a Ranalian flower is shown, a complication being the strong tendency towards pseudo-phyllospory of the androecial elements. This leads to a transformation of the more or less simplified anthocorm shown at left into the still helical Ranalian flower with follicular carpels and peculiar 'leafy' microsporangiate organs (the thecae secondarily separated). In the second diagram petals, etc., are omitted.

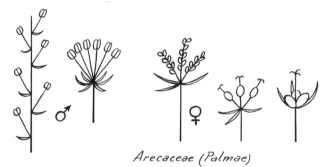

Arecaceae (Palmae)

FIGURE 19C. As Figs. 19A and 19B, but showing a different semophyletic pathway in that the gynoecium did not attain the level of the carpellate pistil (compare Fig. 18). In this putative derivation of the 'flowers' of the Arecales (palms), the oligomerisation of the gonoclads is the prevailing trend. The gonoclads of the unisexual anthocorms became reduced to single stamens which retained their androclad bracts and became cyclic, and to single ovuliferous elements of cupular derivation which likewise retained their bracts, in the male and the female, respectively. A complication is that the whorl of (usually three) reduced gynoclads has formed a common 'stylar' organ.

many features in common with the Pandanales but the early loss of the gynoecial bracts during the early evolution of the latter arrested their gynoecial morphology and isolated the order, whereas related but unspecialised forms which retained these bracts could give rise to various other lines leading to recent groups of the Monocotyledons (Arales, Liliiflorae, Scitamineae, Cyperales, Restionales-Poales, Arecales, etc.) in which additional divergent evolutionary trends and specialisations decided the individual pathways of their subsequent evolution.

Several tentative typological derivations based on the primitive anthocorm are added in the form of series of diagrams (see Figs. 15 and 19A, B, C) as illustrations of early divergencies, initial specialisations, frequent parallelisms and extreme simplifications, which characterise the complicated phylogenetic history of the Angiosperms.

19

A Phylogenetic Approach
to the Morphology
of Seeds and Fruits

In colloquial language, the word 'seed' is generally employed for various products of sexual reproduction which contain the young sporophyte, but laymen do not generally distinguish 'seeds' from 'fruits'. When these words came to be employed as phytomorphological (descriptive) terms, they were defined in an Angiosperm-centred fashion, *viz.*, a 'seed' as the derivative of a *fertilised* ovule (*i.e.*, of the nucellus and the integuments) and a 'fruit' as the mature stage of an ovary or a pistil (*i.e.*, of a 'carpel' or several associated 'carpels'). It became customary to speak of the 'seeds' of Conifers, Cycads and other Gymnosperms, which is in order because they are indeed the mature derivatives of ovules and fundamentally homologous with the seeds of the Angiosperms. However, the term 'seed' came to be used in too wide a sense and the original circumscription certainly does not include various categories of female reproductive organs of cormophytic groups which have at one time or other been called 'seeds'. The best example is provided by the name 'Pteridospermae' or 'seed ferns' given to a group of plants after the discovery of the organic connection between their cupulated ovules and their fern-like fronds. The names suggest that these plants produce 'seeds', but attached cupules never contain a young sporophyte (nor has an embryo ever been found in detached fossil cupules, for that matter). The question arises how far the concept 'seed' can be stretched, because one either has to draw a line somewhere if the original definition is to be maintained, or must give the term 'seed' a completely different circumscription. This is another example of the discrepancies and inconsistencies resulting from the application of Angiosperm-centred semantics to the more primitive conditions in early (primitive) Tracheophytes. For various reasons, it is advisable to retain the term 'seed' in its original meaning (*i.e.*, a derivative of an ovule *containing the young sporophyte*) and to exclude all other categories of megasporangiate reproductive organs retained for some time on the sporophyte after the development of the sporangium, irrespective of their homology with ovules or seeds. This restriction would permit the use of the term 'seed' in the customary sense in the majority of the higher cormophytic groups and yet not interfere with the inquiry into the semophyletic connections between 'seeds' and other categories of retained ovules (*presemina*). Admittedly, such a distinction cuts through the semophylesis of the seed, but this level between *presemen* and seed coincides more or less with the natural groups among the Spermatophyta and thus all living Coniferophylina and Higher Cycadopsids can be called spermatophytes, whereas all pteridospermous groups are excluded from the true seed plants. Such a practical subdivision also corresponds with the established classification of the Higher Plants as '*Spermatophyta*' and is to be preferred to EMBERGER's quite arbitrary distinction between '*préphanérogames*' and '*phanérogames*' (which has been discussed in another chapter in connection with the confusion of 'lines' and 'levels'; see p. 65).

The term 'fruit' has traditionally been reserved for the seed-containing structures of the Angiosperms, defined as the mature derivatives of the 'pistils'. According to the classical interpretation, all gynoecia of the Angiosperms are carpellate and this implies the homology of all pistils and of all fruits, so that the typological derivation of all kinds of fruits from a common basic type (the apocarpous Ranalian follicle) was natural. The classification of the fruits as berries, drupes, nuts, achenes, capsules, etc., is to a large extent based on the same assumption. A berry and a drupe, for instance, are by definition distinguished by the consistency of the derivatives of the innermost layers of carpellary tissue.

In the discussion of the semophyleses of the gynoecia in Chapters 15 and 16, the occurrence of at least two non-homologous categories of pistils has been demonstrated. It follows that at least two corresponding types of mature derivatives must be distinguished, one of which is fully homologous with the cupulate *seeds* of cycadeoid and chlamydospermous Gymnosperms. I shall return to this point later, and first give an outline of the evolution of the seed.

The seed is the ultimate stage of a long semophylesis, in the Higher Cormophyta, that began with the advent of heterospory. Heterospory originated in several main evolutionary lines of the Cormophyta (also in Lycophyta and Sphenophyta), but it did not always lead to the formation of a seed. This requires not only the retention of the megaspore on the mother plant, but, if we associate the term 'seed' with the presence of the young sporophyte of the next generation, also the development of the female gametophyte and the fertilisation of the female gametes (egg cells) within the retained megasporangium. If the megasporangium is shed before a zygotic nucleus has been formed, it cannot be called a 'seed'. The development of the seed was perhaps an exclusive feature in the evolution of gymnospermous groups; at any rate, the recent hetero-sporous Pteridophytes (*e.g., Isoëtes, Selaginella*) never did evolve beyond the initial phases of the evolution of the seed, and it is doubtful whether the macrosporangia of their fossil relatives such as *Lepidocarpon* ever attained the semophyletic level of a true seed. Still, the heterosporous Lower Cormophyta give us an insight into the development of the early stages in the evolution of the seed. At the early 'pteridophytic' level of reproduction, the megaspores were shed like the microspores before germination and the subsequent growth of the gametophyte, the fertilisa-tion process and the development of the young sporophyte proceeded quite independently of the mother plant. The first innovation, perhaps already occurring in the Progymnosperms and also known in some species of *Selaginella*, is the 'precocious' germination of the megaspore and the germination of the microspores in the vicinity of the megasporangium. This phase could extend without an essential change till it included the complete development of the female gametophyte before the megaspore is shed. The next important step was the development of certain features which facilitated the catching or trapping of the microspores. The ovules of the early-Cycadofilicales with long plumose extensions of the salpinx must have caught the microspores in very much the same way as ane-mophilous Angiosperms such as grasses collect pollen grains on their stigmata (a beautiful example of a convergence or 'analogy'!). Younger representatives of this group of seed ferns had a 'pollen chamber', in which the microspores collected after having been caught by the lobes of the integument forming the microphyle. As far as can be ascertained the cupules or their contents were shed before the microspores had germinated, the fertilisation process and the development of the young Sporophyte occurring after the ovules had become detached from the mother plant, so that the ovules did not attain the level of a seed. (I may add that the conditions are still very much the same in *Salvinia, Azolla,* and the Marsileales which I consider to be dwarfed surviving Pterido-spermae: the fertilisation takes place after the shedding of the cupulate megaspores!) The trapping of the microspores had the enormous advan-tage that as soon as the microspores could germinate the fertilisation was almost assured. The male gametophyte was already reduced in that an extensive male prothallus was not formed. In the Carboniferous terres-trial forms, the fertilisation presumably did not follow unless the condi-tions were favourable: for instance, the accumulation of moisture in the pollen chamber (after rain had fallen and soaked the presemen lying on the ground, etc.). From this level it is only a small step to the next phase, in which the antherozoids were released from the microspores in the pollen chamber before the ovules were shed, but it is probable that external conditions had to be favourable (because a sufficient quantity of a watery liquid had to collect in the pollen chamber). The situation found in *Ginkgo* and the living Cycads is again somewhat more advanced, because the development of the male gametophyte and the subsequent activities of the antherozoids have become completely independent of the uncertain external climatic conditions by the formation of a tubular outgrowth of the germinating microspore, which becomes attached to the side of the pollen chamber and is obviously a feeding organ (haustorium) that liberates its isotonous liquid contents together with the antherozoids. The gametic fusion takes places before the ovule is shed (not invariably so in *Ginkgo biloba,* which is transitional in this respect) and the evolution of the seed has begun.

Three tendencies contributed towards the steadily increasing efficiency of the process of reproduction, trends which were operative in more than one of the parallel phylogenetic lineages of the Higher Cormophyta. The retention of the megaspore and the 'parasitic' development of the male gametophyte resulted in a progressive reduction of the gametophytes which became extreme in the Angiosperms. The second trend was the change-over to siphonogamy, which can be visualised as a gradual semophyletic lengthening of the free portion of the feeding haustorium of the male gametophyte until it came into contact with the female gametophyte and eventually penetrated it before discharging its contents. This must have been concomitant with the degeneration of the two antherozoids into free nuclei. The third tendency is the progressive development of the embryo before the shedding of the seed, culminating in the typical angiospermous seed that has a resting stage between the complete development of the embryo and the beginning of germination.

A seed is always covered by an integument, because the spermatophytic level of evolution was reached only after the megaspore had evolved into an ovule. In Chapter 15 the development of additional protective covers of the ovule has been discussed. As a derivative of these covering layers the seed coat or testa includes the second integument, sometimes in the form of an epimatium (Podocarpaceae) or of various other homologues referred to as 'arils' (*Ginkgo, Taxus*), a 'sarcotesta' (Cycadales), or seed wings (Pinales). In the seed ferns the presemina were enclosed in a cupule which later became the chlamys or fleshy 'exotesta' (Chlamydospermae and other groups of Bennettitalean affinities) and is in fact the prototype of the true arils of the Angiosperms. The increased 'protection' of the ovule often resulted in a more or less complete isolation of the female gametophyte from the outside world, so that the microspores no longer had free access to the pollen chamber. In several forms, the inner integument is prolonged into a beak-like micropylar projection with a very narrow central canal and the pollen grains germinate at the outside on or near the exostomium of the micropyle, sending out long pollen tubes extending to the nucellus. The protection of the ovules, a process often explained as an adaptation to entomophily, thus indirectly furthered the development of siphonogamy, the advent of which created a new situation. At the level of zoidiogamy, only a single antherozoid normally enters the neck of an archegonium (Cycadales!), and in each female gametophyte only one male gametic nucleus fuses with an egg cell to form an embryo; but siphonogamy results in the release of *two* male generative nuclei inside the female gametophyte. These two gametic nuclei are to all intents and purposes identical, so that each of them has the capability of combining with a female gametic nucleus to form a zygotic product. The embryology of *Gnetum* demonstrates this equivalence, because several potential egg cells are formed in the embryo sac which can be fertilised by either generative male nucleus of a pair released from the same pollen tube. In Cycadales, *Gnetum* and a few other cases normally only one fertilised egg cell per ovule develops into an embryo, irrespective of the number of gametic fusions. The other zygotic nucleus soon degenerates or performs only a few mitotic divisions before degenerating, apparently because the developing embryo inhibits the formation of additional embryos. In the Coniferales and the Angiosperms there is only one egg cell, which requires only one of the two male generative nuclei, so that the other one is 'redundant'; but the second male nucleus retains its disposition to fuse with a female gametic nucleus. If there is no second egg nucleus but more or less equivalent female nuclei are present in the vicinity, a second fusion is likely to occur, so that the conditions favouring a double fertilisation are the almost inevitable consequence of the process known as siphonogamy: in other words, double fertilisation is a semophyletic level that could in principle be attained in every phylogenetic lineage in which siphonogamy became established and, accordingly, arose polyrheithrically rather than monophyletically.

The double fertilisation must not be seen as a process that invariably leads to the formation of a secondary endosperm; on the contrary, the inhibition of the development of a second embryo in Cycadales and Chlamydospermae suggests that a second zygotic nucleus would not have much chance of survival. In the Coniferales a secondary endosperm has never been found, and when an endosperm is formed in an Angiosperm —which is not always the case—it is usually formed by the 'triple' fusion of the two haploid female polar nuclei and the second male generative nucleus.

These two situations may be indicative of a possible physicochemical (perhaps a serological) counter-balancing of the inhibition of the potential second embryo. The growing embryo does not inhibit the development of the maternal diploid tissues of the ovule such as the integuments and the aril, so that conceivably its antagonistic effect on a triploid nucleus which is two-thirds maternal and only one-third paternal would be less than its effect on a diploid zygote which is 50 per cent paternal. This speculation might be given a phylogenetic interpretation by assuming that the development of a secondary endosperm could begin only after a singular fusion of two female haploid nuclei (and the fusions between haploid female nuclei in the embryo sacs of some Chlamydospermae may be significant in this connection), but in view of the many alternative cases—the polar nucleus may be triploid to polyploid—this remains conjectural.

That the endosperm, as a potential second embryo, is a storage tissue is not surprising. Since it is a heterozygous derivative of parental haploid nuclei, an eventual plasmatic incompatibility between the maternal feeding tissues and the developing embryo is thus partly suppressed, so that the advent of the secondary endosperm had survival value and may have been a positive selective factor.

Much has been written about the relation between the characters of the endosperm and the phylogeny of the angiospermous taxa. Such terms as 'nuclear' or 'cellular' endosperm, 'helobial endosperm', 'Farinosae', etc., suggest that the mode of development and the consistency of the endosperm have a considerable bearing on the classification of angiospermous taxa. This may be true, but it is not at all certain whether there is only one original (primitive) condition (say, the 'nuclear' type of endosperm development). The development of the secondary endosperm is so diverse, that the phylogenetic relations are not clear.

The outer layers forming an integral part of the seed, the testa, are derived from tegumentary covering layers of the ovules, but the outer integument is not always, and the chlamys or aril but rarely, incorporated in the seed skin. If we base the nomenclature of the cycadopsid seeds and their outer coverings on the basic homologies of the ovules and their protective organs, the mature derivatives of certain categories of angiospermous pistils which are the homologues of arillate ovules must be called *seeds* and not fruits (*e.g.*, in Piperaceae, Juglandaceae, Urticales, Cyperaceae). The lateral coalescence of these seeds into complexes produces *synspermia* (not syncarpia!) or 'phalanges' (*Sararanga, Pandanus p.p.*, many palms, Restionaceae, Flagellariaceae).

There is no reason to refer to the mature derivatives of all forms of carpellate ovaries by any other name than the traditional term 'fruits'.

The typological classification of these fruits as follicles, legumes, berries, drupes, achenes, capsules, etc., can also be maintained, as long as it is understood that several of these categories represent parallelisms (*e.g.*, the capsules of Caryophyllaceae and Primulaceae, the berries and achenes of centrospermous groups such as Basellaceae, Chenopodiaceae, Amaranthaceae and those of Rosaceae, Ranunculaceae, Berberidaceae). The carpellate ovaries are not all derived from a single (common) prototype of pistil, so that there are certainly no grounds on which to 'derive' all fruits from the same prototype (such as the apocarpous Ranalian follicle). Apocarpy and syncarpy are often *alternative* conditions (cf. Rosales, Annonaceae!) and not necessarily indicative of the relative advancement of taxa exhibiting these features. The ovaries of many Centrospermae, Liliiflorae, Guttiferae and Cistales, for instance, originated as 'coenocarpous' structures and not as aggregates of apocarpous follicles. The question arises what the most primitive types of fruits have looked like. In his Durian theory, CORNER tries to make out a case for the fleshy dehiscent follicle with arillate seeds. I cannot subscribe to this idea, because (1) not all fruits are phylogenetically derived from follicles, and (2) several divergent adaptive evolutionary trends in the method of dispersal of the seeds must have been operative from the beginning, *i.e.*, before the advent of the carpel.

The oldest dispersal unit was the *seed* or the *cupule*, not the fruit. In the Higher Cycadopsids the seeds are originally chlamydote (arillate), and in the more primitive forms this outer layer was *usually* fleshy and zoochorous. This type of zoochorous seed is still found among the Chlamydospermae (*Gnetum*), Piperales (*e.g. Piper*), in some primitive Cyperales, in some Pandanaceae and Palmae, etc., but soon other trends developed, the outer layer becoming fibrous (*Pandanus*, palms) or dry (Cyperaceae, Restionaceae, some Urticales, etc.). In the latter case the seeds are phenetically very similar to achenes or one-seeded capsules (and in their dispersal often behave like fruits, *e.g.*, by dehiscing, as in Urticaceae). *Welwitschia* and *Ulmus* demonstrate that anemochory was another early trend which is associated with the development of a dry, winged seed-coat. These trends must have been well established before closed carpellate ovaries developed and the angiospermous fruits came into being. It is not very likely that seeds already adapted to anemochory required a fleshy and edible pericarp as an aid in dispersal, so that it is plausible that the fruits of, *e.g.*, Salicaceae must have had dry-walled (and dehiscent) prototypes. The seeds and not the fruits being the primary dispersal units, seeds provided with an edible aril did not need a fleshy pericarp either, as long as the fruit remained dehiscent. The dehiscent non-fleshy fruit seems to be the primitive condition, provided the bracts of the gynoclads were in the pre-carpellary phases of the evolution of the gynoecia not already adapted to a function in dispersal in the same way as calyces, bracts, receptacles, pedicels and other 'extraneous' organs are often incorporated in functional dispersal units containing one to several fruits (cf. *Tilia*, Dipterocarpaceae, *Anacardium*, *Fragaria*, etc.). Phylogenetically, the dehiscent fruit is the most likely primitive type, considering that the walls of carpellate ovaries are derivatives of laminose organs which gradually became completely closed and that the commissures must have remained preformed weak spots that became lines of dehiscence. The first fruits must already have been diverse in that the seeds were adapted to either zoochory or anemochory. The subsequent radiating evolution of the fruits, and the seeds they contain, in various directions has resulted in numerous derived types of fruits. One trend is the change-over from a dehiscent to an indehiscent fruit; in such cases the fruit becomes the dispersal unit, often requiring a 'transference of function' from the seed-coat to the pericarp, so that the latter became fleshy and zoochorous, or winged, dry and anemochorous, etc. The development of fleshy berries out of dry capsules (or *vice versa*) is still proceeding, and such transitions can be found in several unrelated families: Cucurbitaceae, Gentianaceae, Campanulaceae, Convolvulaceae, Liliaceae. Many parallelisms must be anticipated, and the typological classification of the fruits, no matter how useful for descriptive purposes, does not permit a phylogenetic interpretation of the relationships of the taxa bearing the same or a different form of fruit without corroborative evidence from other sources.

The phylogenetic approach to the morphology of fruits (and seeds) outlined in this chapter opens up new perspectives and clearly demonstrates that certain 'established' points of view should be reconsidered. The most sweeping innovation is the interpretation of certain traditional 'one-seeded fruits' as arillate *seeds*, but the rejection of the conventional derivation of all fruits from a single basic archetype (the Ranalian follicle) is no less fundamental. That taxa with syncarpous (or coenocarpous) fruits are derived from apocarpous ancestral forms has been axiomatic, but there are several alternative primitive categories of fruits including, apart from the apocarpous follicle, primary syncarpous or coenocarpous types. The recognition of the occurrence of alternative 'primitive' conditions relieves the systematist of the traditional obligation, to derive all morphological types of fruits (and all angiospermous taxa) from the same stereotyped prototype that has hitherto cramped his style, and may enable him to see the phylogenetic relationships in their true perspective.

20

Gymnosperms and
Primitive Angiosperms—
A Retrospective View

The phylogenetic relations between the cycadopsid Gymnosperms and the Angiosperms have repeatedly been discussed in preceding chapters, but a general summary does not seem to be out of place. The relationships can be expressed in the form of the following basic assumptions and deductions:

(1) The Angiosperms have descended polyphyletically (polyrheithrically) by way of a number of parallel evolutionary lines which were most probably already separated in early Mesozoic epochs and, in the initial phases of their independent evolution, still at the 'gymnospermous' (*i.e.*, chlamydospermous-Bennettitalean) level of organisation.

(2) Certain 'Angiosperm trends', barely, or not at all developed in the basic taxa of these parallel lineages, evolved as parallelisms in all these phylogenetic lines; but not all trends developed in every genealogical sequence leading to a recent angiospermous terminal group.

(3) These 'Angiosperm trends' include morphological (phenetic), anatomical, embryological, palynological and biological (ecological) adaptations and differentiations, some or perhaps all of which can be seen as a group of correlative evolutionary phenomena that can be ascribed to a single main causal force evoking morphological adaptation through positive selective tendencies. This fundamental cause can be defined as 'angiospermy' (angiovuly), *i.e.*, protection of the ovules, as a reaction to the advent of entomophily (GRANT 1950, STEBBINS 1951).

(4) The initial adaptations involved the reproductive organs and the first morphological indications of the evolution towards angiospermy are likely to be discernible in the reproductory regions, the vegetative organs lagging behind.

(5) General typological and phylogenetic considerations indicate that the female (macrosporangiate) organs of the cycadopsid prototypes had

at least attained the level of an ovule, fertilised by means of siphonogamy soon resulting in an early stage of double fertilisation; the ovules, typically bitegmic and provided by a third enveloping layer of cupular derivation (a chlamys or aril, or its equivalent), were not borne singly but in groups (trusses) based on a common supporting axis (gynoclad) subtended by a bract or stegophyll.

(6) The male (microsporangiate) organs were similarly organised in compound structures (androclads), likewise subtended by a bract, the ultimate branches bearing terminal synangia of typically four sporangia.

(7) The gonoclads (gynoclads and androclads) were inserted on a common supporting axis (which may, in addition, have borne some sterile leafy organs in its lowermost portion) to form strobiloid complex structures (anthocorms); these anthocorms were either exclusively male or female (corresponding with 'monoecy' or 'dioecy'), or bisexual (amphisporangiate); in the latter case the gynoclads usually occupying the most distal portion and the androclads forming a zone below them.

(8) Anthocorms developed into the functional reproductive units of the Angiosperms, but not always into the conventional 'flowers', depending on the homology of the functional or typologically postulated 'flower' with a whole anthocorm or with a portion of it (a gynoclad, sometimes only a solitary ovule), so that the term 'flower' must be re-defined and is best circumscribed as the derivative (homologue) of an entire anthocorm.

(9) During the phylogeny of the Angiosperms several reductions and changes took place in the anthocorm and its constituting elements, the most general tendency being a progressive oligomerisation of the number of gonoclads, but reductions in the number of ovules and androsynangia ('stamens') also frequently took place.

(10) A characteristic evolutionary process in several, but not in all, of the phylogenetic lines leading to the recent Angiosperms was the development of carpellate gynoecia (pistils) by the semophyletic combination of one to several gynoclads with the corresponding subtending bracts; this must have come about by a repetitive (homoplastic or parallel) evolution of a certain archetype to a phenetically identical end phase.

(11) The androecia underwent various reductions, but the androclads retained their individuality as a rule (except in cases of extreme pseudophyllospory of the androclad-bract unit, as in Magnoliales, and in some advanced, e.g., gynandrous, taxa).

(12) The bracts subtending the gynoclads are not so often retained as individual entities (see item 10), but those axillant to the androclads usually remain distinct and, often together with the sterile phyllomes of the proximal region of the anthocorm, provide the elements constituting the perianth of the 'flowers'.

(13) The various tendencies towards semophyletic changes in the floral region developed heterobathmically in the various angiospermous lineages, so that one trend prevailed or preceded others, and, for instance, a progressive oligomerisation of the number of ovules or male synangia per gonoclad may have begun early (leading to the seemingly very simple 'flowers' of, e.g., Piperaceae and Cyperaceae), whereas in other lineages numerous ovules and androsynangia were retained (Dilleniaceae, Flacourtiaceae, Clusiaceae, Myrtales, Velloziaceae, etc.).

(14) A universal trend, the angiospermy (i.e., the development of the gynoecia into closed structures) involved various morphologically different organs, the enveloping layer being formed by ovular coats (the outer integument, the chlamys, or both, by an over-arching and, eventually, closure of the micropylar end), by the conduplication and closure of a single pseudo-sporophyll (many Polycarpicae, Leguminosae) or by the association of several gynoclad-bract units (Parietales-Guttiferae, Centrospermae, Liliales, Scitamineae, etc.).

(15) The encasing of the ovules in an enveloping structure required the formation of stigmatic (pollen-catching) areas which are derivatives of the ovular coats in the non-carpellate gynoecia, of the carpel wall in many monocarpellate pistils (e.g., in Magnoliales), or of both the ovular coverings and the sterile ovary wall (probably of frequent occurrence), eventual stylar extensions of the pistils being formed out of the same elements.

An angiospermous taxon deserves the qualification of a 'primitive' group only if it fulfils at least one of two requisites, viz., either a retention of several ancient stages of the more specific characteristics of the Angiosperms, its other features being more or less advanced, or a fairly general advancement of most characters which still enables the recognition of the taxon in question as the intermediate between some basic group and a considerable number of more advanced derivatives. These two conditions are not always clearly distinguishable, but archaic forms are often advanced (or even over-specialised!) in certain respects and do not correspond with the 'general' morphological pattern of the group of which they are an offshoot (in other words, they are not representative of the group of progenitors as a whole), whereas the second, not so specialized, group of primitive forms reflects early semophyletic stages of most of its characters and is a more or less advanced edition of the representative elements of the ancestral taxon. The plants belonging to the first category survived because they became morphologically static in some respects and their type thus became 'frozen'. The less specialised groups, which include the original near-allies of the archaic forms exhibiting early specialisation, evolved much more gradually in various directions, their original primitive morphological pattern altering slowly, but con-

sistently, so that ultimately they became so far removed from the archetype that the latter 'disappeared' and became 'extinct'. The extreme
cases of the first category are much more readily recognised as 'ancient'
types, because they have retained some striking primitive features so
that one is tempted to conclude that all or nearly all their morphological
characteristics are 'primitive', *i.e.*, representative of their ancestral group.
The tacit identification of the morphology of the archetypes with that
of supposedly 'primitive' contemporary orders has introduced some erroneous ideas regarding the morphology of the Protangiosperms. The
pseudanthium theory of DELPINO-WETTSTEIN postulates that these pro-
taxa were related to the recent Chlamydospermae (which indeed holds
true for some major taxa of the Angiosperms!), but the fundamental misconception was the assumption that the floral morphology of such forms
as *Ephedra* and *Casuarina* is representative of the group, or groups, of
which they are the survivors. The gonoclads of the Chlamydosperms are
very much reduced and it is of course impossible to derive a bundle of
co-axial petal-opposed stamens of, *e.g.*, *Paeonia*, Dilleniaceae and Guttiferae, or a pistil with a considerable number of carpels and many ovules
per carpel, from reduced gnetalian genitalia. A common group of progenitors of recent Chlamydosperms and Angiosperms must have had many
gonoclads per anthocorm, and the gonoclads must have borne numerous
cupulate ovules or androsynangia. This is corroborated by the conditions
in the Piperales (which are, in my opinion, the closest living relatives of
the Chlamydospermae), the Piperaceae (Peperomiaceae) and Chloranthaceae exhibiting an advanced oligomerisation of the number of gonoclads, ovules and male synangia, to a single ovule and one to three
(occasionally up to ten) solitary stamens per anthocorm, respectively,
but the Saururaceae possessing plurilocular pistils with at least two
ovules per locule surrounded, usually, by six to eight stamens. The
anthocorms of the Saururaceae already approach the flower type as
found in, *e.g.*, Ranunculales, and but for the absence of a perianth are
in fact rather typical (cyclically organised) angiospermous flowers. The
Saururaceae are advanced over the other Piperales in some respects, such
as the organisation of the gynoecia, but the oligomerisation of the genitalia
(also a derived condition!) has not progressed as far as in the other
piperalean families. The Chloranthaceae are undoubtedly the most primitive in their anatomical and embryological features (see MEEUSE 1963b),
but the specialisation of the anthocorms made them a static archaic offshoot. The Saururaceae did not become so specialised, and conceivably
this enabled them to progress substantially in the general direction of the
Polycarpicae.

The same error of judgment was made in the euanthium hypothesis,
in which the Polycarpicae (and especially the Magnoliales) are presumed to be the most primitive living Angiosperms. The androclads in

the majority of the Magnoliales are invariably reduced to single bithecate anthers, each often merged with its bract to become pseudo-phyllosporous, so that they must have shown these trends at an early stage of evolution, precluding the direct descent of forms with non-phyllosporous pluristaminate androecia (*Paeonia*, Dilleniaceae, Parietales, Guttiferae and several other dialypetalous groups) from the magnolialean type. The gonoclads of the Magnoliales became monocarpellate ovaries by a neotenic development leading to pseudo-phyllospory, and it is equally unimaginable that forms with non-carpellate ovaries (Urticales, Juglandaceae) or even such forms as Laurales, *Nymphaea* and Centrospermae could have descended from archetypes with a magnoliaceous floral morphology. All things considered, the Magnoliales seem to represent an ancient surviving offshoot with early specialisations and hence a rigid morphology. In accordance with the general rule, they retained some very primitive features (such as homoxylous wood and monosulcate pollen grains). The proto-Ranalians, at least the forms with monocolpate pollen grains, may have resembled the Magnoliales in such archaic features as a primitive wood anatomy, arillate seeds, a testa of primitive construction, etc., but they must have had a still unspecialised morphology of the reproductive region, *i.e.*, primitive anthocorms. The more successful angiospermous groups, *i.e.*, Ranunculales, Dialypetalae (and Sympetalae), Centrospermae, and many Monocots, are not so specialised (they do not exhibit pseudo-phyllospory of the androecia, there are often more ovules per carpel than is average in the Magnoliales, and the pseudophyllospory of the gynoecia is usually not so pronounced), and the only arborescent order possibly derived from the Polycarpicae, the Hamamelidales, cannot be a descendant of the Magnoliales either, because they belong to the group with basically tricolpate pollen. This leaves virtually not a single group which could be a derivative of the Magnoliales!

Typological and other evidence, *e.g.*, palynological and anatomical, indicates so consistently that the majority of the old Dialypetalae (minus most of the Polycarpicae, Hamamelidales and Centrospermae) and, indirectly, all Sympetalae, have descended from a 'plexus' which must have given rise to a number of more or less primitive families, some of which are not clearly separable even at the present-day level (and have been referred to different orders in the various systems of classification), that one can accept this conclusion in a phylogenetic sense. An analysis of probably the most primitive families of the plexus, on the understanding that these must have gynoclads (*i.e.*, carpels) with numerous ovules and androclads bearing many stamens, arillate seeds, dehiscent fruits, a more primitive wood anatomy, etc., enables us to select all characters that are common to these taxa (which, in this case, almost certainly represent truly homologous conditions) and are, therefore, also the original features inherited from their common ancestral group. If it is typologically impossible to select a single common character, at least two or three alternative conditions can be indicated. The assembly of primitive characters must represent an approximation of the diagnosis of the common ancestral taxon. As examples of the families that can be used to draw up such a 'synthetic' diagnosis, the Dilleniaceae, Crossosomataceae, Actinidiaceae, Dipterocarpaceae, Theaceae, Ochnaceae, Flacourtiaceae, Clusiaceae, Elaeocarpaceae, Cunoniaceae, 'Brexiaceae', and 'Davideaceae' can be mentioned. The probable prototaxon must have had the following characteristics: The habit must have been that of a woody plant, but not necessarily of a large tree, more probably that of a shrub or small tree (occasionally of a rhizomatous plant of low stature).

No cogent reasons can be cited for the assumption that this group was monocaul and pachycaul (as CORNER postulated for the primitive dicotyledonous tree in his 'Durian Theory' of 1949). The leaves must have been petiolate, presumably showing the initial phase of the formation of stipules (petiolar wings as in, *e.g.*, many Dilleniaceae), both simple and compound types being present, but the simple ones betraying their origin from a compound type by being palmately lobed or penninerved, with parallel veins ending in the tips of the sharply serrate margin (craspedodromous); the phyllotaxis must have been alternate or perhaps sometimes decussate. The anthocorms were predominantly bisexual, solitary or aggregated in a terminal paniculate compound structure, and exhibited early trends towards cyclisation, pentamery and the formation of a corolla out of the bracts supporting the androclads. The androclads, soon in a single or double whorl of five (rarely four), bore numerous stamens and must originally have been quite free from the floral axis, but later tended to fuse with the torus, only the stamens remaining free (and appearing as individual stamens). The gynoclads, mostly arranged in a single whorl and initially free from the floral axis, bore numerous bitegmic arillate ovules, the ovules of each gonoclad forming a common (phalangial) stylar extension; their bracts, originally free, tended to fuse at the base and to form centripetal plate-like extensions (the septa) usually alternating with the gynoclads; gynoclads and bracts later showed different trends, *viz.*, (1) pseudo-phyllospory of each individual gynoclad-bract unit, (2) the adnation of the gynoclads to the floral axis and lateral concrescence of the bracts to form a loculate pistil with axile placentation, and (3) the combination of the assembly of gynoclads with their bracts, leading to the formation of various types of carpellate gynoecia with 'parietal' placentation. The seeds were almost certainly arillate and had some endosperm.

The picture thus obtained—in fact already a rough 'diagnosis' of the putative progenitors—does not particularly strike us as unusual; on the contrary, it is anything but exciting and (provided the ovaries are closed or nearly so) could still apply to representatives of several recent families (such as Dilleniaceae). The reconstruction, 'the extrapolation into the past', has actually yielded an example of the *unspecialised*, unspectacular primitive group which is the veritable but unobtrusive prototaxon, in contrast to the more or less specialised and much more 'obviously' archaic forms such as Piperales, Magnoliales (and Chlamydospermae) discussed in the beginning of this chapter. If one attempts a further extrapolation of the lineage of the reconstructed prototaxon, one arrives at another general type of plant with, for instance, helically arranged gonoclads, but this cannot be very much different from the Mesozoic protangiospermous group of plants of chlamydospermous-Bennettitalean alliance with primitive anthocorms postulated on the basis of other considerations (see, *e.g.*, Chapter 10) and *de facto* represented by the Pentoxylales, which are, however, ancestral to the Monocotyledons. The typological derivation of the reproductive structures of all Angiosperms from such protangiospermous forms is possible, but this provides also a phylogenetic argument in that the relationships among the Rosiflorae-Parietales-Guttiferae plexus, the Magnoliales, Ranunculales and other Polycarpicae, the Piperales, the Amentiflorae and the Centrospermae must root in an early and probably still 'gymnospermous' common ancestry.

The chances of finding modern representatives of the hypothetical basic group of the large dialypetalous plexus are slender, because this group has 'disappeared' as the result of a slow but continuous adaptive radiation. Still, there is at least one example that comes amazingly close to it. Several of the more primitive Dilleniaceae have free or semi-apocarpous carpels that are open at the distal end. If one assumes that the placental axis (gynoclad) is not adnate to the central axis or the gynoclad bract (a valvular segment of the ovary wall) but in a free axillary position in respect of its stegophyll and that the pluristaminate androclads (sometimes partially free in some Dilleniaceae) are completely free from the torus, one arrives at the condition postulated for the reconstructed prototype. The pinnate or bipinnate leaves of the acaulescent Acrotremeae are also suggestive of a fern-frond, *i.e.*, a pteridospermous-cycadopsid phyllome, and of one of the possible habit forms of the 'reconstructed' group of Protodicots. The Dilleniaceae are apparently not very far advanced beyond this deduced ancestral stage (which is corroborated by anatomical evidence: wood anatomy and nodal structure) and a close scrutiny of this family or related taxa, perhaps a lucky discovery, may eventually provide us with an even closer approximation of the postulated ancient group. At any rate, the morphology of these suggested early Angiosperms (and the Dilleniaceae) is a far cry from the conventional idea that *the* protangiosperms resembled the Magnoliales (Polycarpicae, Ranales), and it is high time that these traditionalisms were relegated to the waste-paper basket.

The position of other groups of Dicotyledons is not at all clear, but it is highly probable that at one time a differentiation of primitive monosulcate sporomorphs into porate and tricolpate types of pollen initiated.

PLATE V. A primitive representative of the dialypetalous Dicots, *Dillenia obovata* (Bl.) Hoogland, bearing flowers on its branches. Gunung Pantjar, western Java. (Photo by C. N. A. DE VOOGD. Courtesy of Flora Malesiana Foundation.)

The key taxon again seems to be the Piperales with their peculiar eurypalyny, not only showing relations with Chlamydosperms on the one side and with monocolpate dicotyledonous groups on the other side, but also exhibiting the inaperturate condition, which, at a pinch, could be interpreted as a relationship with such forms as the Laurales. This would be compatible with the suggestion that a chlamydospermous-piperalian alliance with primitive anthocorms split up into several lines, at least one leading to the monosulcate Magnoliales, one to the tricolpate groups and perhaps additional ones to the porate Monochlamydeae, to the Laurales and to the Aristolochiales. Nevertheless, the position of the Ranunculales-Papaverales, Hamamelidales, Centrospermae, Monochlamydeae, *Nelumbo* and the Nymphaeales in respect of the Magnoliales and of the plexus of the dialypetalous Dicotyledons remains more or less obscure, so that the postulation of more ancient lineages or early offshoots appears to me to be ineluctable. However, a further discussion would be too speculative.

The Monocotyledons are apparently much more homogeneous in that (the Dioscoreales excepted) they can all be typologically related to a group of fossil plants of which the Pentoxylales are representative but already somewhat specialised examples as will be discussed presently. The typological connections between the recent Pandanales and the Jurassic Pentoxylales are so striking that there can be very little doubt that there is a phlogenetic relation between the two (MEEUSE 1961a).

The reason why the archaic Pandanaceae survived must, according to a general rule, be an early specialisation that 'froze' their floral morphology, and, indeed, they show an important reduction already present in the Jurassic *Carnoconites, viz.,* the absence of bracts subtending the gynoclads (the genus *Freycinetia* may be an exception because it has a different gynoecial morphology, but its pistil has never been studied anatomically). The lack of bracts naturally prevented the formation of carpellate ovaries, but it also provides a clue to the origin of the other Monocots. If one assumes that forms have existed which were closely related to the known Pentoxylales but differed from the latter chiefly in the presence of bracts subtending the gonoclads and of bisexual as well as unisexual anthocorms, all other monocotyledonous groups can be typologically derived from such prototypes. Several main trends can be distinguished which are indicative of alternative parallel developments. In the Cyperaceae, the oligomerisation of the gynoclads to a single ovule and of the androclads to a single stamen or triad of stamens produced the scirpoid flower from a bisexual primitive anthocorm by the reduction of the gynoclads and their bracts to a solitary and *de facto* terminal cupulate ovule, and by retention of three or six androclads reduced to stamens typically without subtending bracts (there is frequently no perigone!), as well as the caricoid reproductive structures from a unisexual anthocorm (or a bisexual one with the androclads most distally inserted) by the reduction of the gynoclads to solitary lateral cupulate ovules (the bract typically being retained, sometimes as a utricle), and of the androclads to bracteated triads of stamens, usually several to many gyno- and/or androclads being retained. In other orders (Arecales, Restionales or Flagellariales), the gynoecium became reduced to a

single whorl of usually three and normally fused pistils of cupular derivation, their bracts often being retained as tepals in the unisexual flowers,
and the androecium to typically six (or three) bracteated stamens; a
further reduction led to the one-ovuled gynoecium of the Gramineae.

The Centrolepidaceae provide another very important clue. In such
forms as *Centrolepis* the gynoecia consist of phalanges of one-ovuled
pistils of cupular derivation, supported by a bract, the ovular coverings
sometimes forming an individual and sometimes a common stylar extension. Each anthocorm has retained only one such phalangiate gynoclad
and this is presumably the reason why the Centrolepidaceae survived as
a relict group, because in many other lines such gynoclads must have
occurred in whorls of three (or occasionally two), or in double whorls
of three, and formed carpellate gynoecia, *viz.*, in the Arales, the Helobiae,
the Juncaceae, the Commelinales (Farinosae, partly), the large liliiflorous
assembly of Liliales, Agavales, Haemodorales and derivatives (Iridales,
Orchidales), the Bromeliales and the Scitamineae (Zingiberales). At
least some of these groups must represent ancient lineages and it is not
at all unlikely that at one time a proto-cyperaceous-centrolepidaceous
plexus existed, from which not only the above-mentioned carpellate
orders but also the non-carpellate glumiflorous types developed as the
result of adaptive radiation (the palms apparently represent an independent earlier offshoot with a closer relationship to Pandanales and Cyclanthaceae). The suggested plexus, if one assumes that it included
caulescent to arborescent forms, in fact approaches the postulated near-
pentoxylahan archetype with primitive unreduced anthocorms (*i.e.*, with
bracteated gonoclads) very closely. Caulescent forms, some with dichotomously branched stems and retained secondary growth, are found in
most of its constituting groups: Cyperaceae, Juncaceae, Bromeliales,
Xanthorrhoeaceae, 'Agavales', Velloziaceae, Scitamineae, especially among
primitive representatives *(Dracaena)*, and the caulescent habit is undoubtedly an ancient though not universal feature of the early Monocotyledons.

The Helobiae and the Arales are somewhat incongruous, their morphological and embryological features being rather aberrant, but there is a
general consensus of opinion that they are true Monocots. There are
reasons to assume that these two groups are of remote common origin
and represent early adaptive offshoots of Protomonocots, the Helobiae
showing adaptation to the aquatic habitat and the Araceae to the development of a 'trap-flower' type of inflorescence. Apart from the more generally recognised points of resemblance between the two taxa, the fairly
general absence of the bracts subtending the anthocorms is another and
rather specific common characteristic. It is not at all unlikely that they
(with Cyclanthaceae and Arecaceae) had a common origin from a

group close to the Propandanales (but with bracteated gonoclads!). The gynoecia of the Arales and the Helobiae are apparently consistently carpellate and primarily compounded of several (often three, four or six) pluri-ovulate gynoclad-bract units (but frequently secondarily reduced, the ovules often to few per pistil), and their androclads are usually reduced to three or four solitary stamens (or to two whorls of three), the bracts forming the perianth in male and bisexual flowers.

Two conclusions can be drawn from the foregoing deductions. The salient typological relation between all angiospermous groups and primitive types of cycadopsid plants with primitive anthocorms (which, I believe, can be given a phylogenetic interpretation) indicates a number of possible direct and indirect homologies of taxa and of reproductive regions and also a number of alternative conditions and parallelisms. Of the many examples, only the frequently suggested relationships between Juncaceae and Cyperaceae will be discussed. These two families have perhaps some affinities, because, among other things, their palynological characteristics (pollen tetrads in the Juncaceae, cryptotetrads or pseudomonads in the Cyperaceae) point to a possible propinquity of origin, but, *e.g.*, the gynoecial morphology is altogether different. The standard approach has been to derive Cyperaceae from (proto-) Juncaceae or *vice versa*, but such attempts are rather inane. The common ancestral group, as we have seen, was a non-carpellate form (the Cyperaceae are still non-carpellate) with many ovules per gynoclad (a primary characteristic of the Juncaceae). Divergent trends, *viz.*, an oligomerisation of the number of ovules in the non-carpellate gynoecia in the lineage of the Cyperaceae and the origin of the carpellate pluri-ovulate pistil during the evolution of the Juncaceae, resulted in two alternative parallel phylogenies of approximately the same geological age, neither of which has a more 'advanced' floral morphology than the other. Other characteristics of the progenitors may of course have been retained in both parallel lines of descent and the typical palynological feature may be such an ancient character.

The other corollary is the consistent compatibility of the phylogenetic and typological considerations with the assumption that the Protangiosperms were cycadopsid Gymnosperms with primitive anthocorms. The search for fossil ancestors of the Angiosperms must be directed towards forms with such anthocorms, and even fragmentary records of isolated gonoclads bearing numerous chlamydote bitegmic ovules or numerous male synangia may qualify. There can be very little doubt that such forms as Caytoniales, Corystospermaceae, Pentoxylales and of course the Chlamydospermae exhibit distinct Angiosperm trends, so that conceivably one need only find some more fossils linking these groups to complete the picture.

Bibliography

ABBE, E. C. 1935. Studies in the phylogeny of the Betulaceae. I. Floral and inflorescence anatomy and morphology. *Bot. Gaz.* **97**: 1–67.

———. 1938. *Ibid.* II. Extremes in the range of the variation of floral and inflorescence morphology. *Bot. Gaz.* **99**: 431–446.

ANDREWS, H. N. 1960. Notes on Belgian specimens of *Sporogonites*. *Palaeobotanist* **7**: 85–89.

———. 1961. *Studies in Palaeobotany.* New York and London.

ARBER, A. 1950. *The Natural Philosophy of Plant Form.* Cambridge.

ARBER, E. A. N., and J. PARKIN. 1907. On the origin of Angiosperms. *J. Linn. Soc. Bot.* **38**: 29–44.

———. 1908. Studies on the evolution of the Angiosperms: The relation of the Angiosperms to the Gnetales. *Ann. Bot.* **22**: 489–515.

ARNOLD, C. A. 1947. *An Introduction to Paleobotany.* New York.

———. 1948. Classification of Gymnosperms from the viewpoint of paleobotany. *Bot. Gaz.* **110**: 2–12.

ASAMA, K. 1960. Evolution of the leaf forms through the ages explained by the successive retardation and neoteny. *Sci. Rept. Tohoku Univ. (Sendai). Second Ser.* (Geol.), Spec. Vol. 4 (Hanzawa Mem. Vol.): 252–280.

———. 1962. Evolution of Shansi flora and origin of simple leaf. *Ibid.* Spec. Vol. 5 (Kon'no Mem. Vol.): 247–273.

AXELROD, D. J. 1950. A theory of Angiosperm evolution. *Evolution* **6**: 29–60.

———. 1959. Evolution of the psilophyte paleoflora. *Evolution* **13**(2): 244–275.

———. 1961. How old are the Angiosperms? *Amer. J. Sci.* **259**: 447–459.

BAILEY, I. W. 1954. *Contributions to Plant Anatomy.* Waltham (Mass.).

———. 1956. Nodal anatomy in retrospect. *J. Arnold Arb.* **37**: 269–287.

———. 1957. The potentialities and limitation of wood anatomy in the study of the phylogeny and classification of Angiosperms. *J. Arnold Arb.* **38**: 243–254.

BAILEY, I. W., and B. G. L. SWAMY. 1951. The conduplicate carpel of dicotyledons and its initial trends of specialisation. *Amer. J. Bot.* **38**: 373–379.

BARBER, B. 1961. Resistance by scientists to scientific discovery. *Science* **134** (No. 3479): 596–602.

BARNARD, C. 1957a. Floral histogenesis in the Monocotyledons. I. The Gramineae. *Austr. J. Bot.* **5**: 11–20.

———. 1957b. *Ibid.* II. The Cyperaceae. *Austr. J. Bot.* **5**: 115–128.

———. 1958. *Ibid.* III. The Juncaceae. *Austr. J. Bot.* **6**: 285–298.

———. 1960. *Ibid.* IV. The Liliaceae. *Austr. J. Bot.* **8**: 213–225.

———. 1961. The interpretation of the Angiosperm flower. (Pres. Address, A.N.Z.A.A.S.) *Austr. J. Sci.* **24**: 64–72.

BECK, C. B. 1960. The identity of *Archaeopteris* and *Callixylon*. *Brittonia* **12**: 351–368.

———. 1962. Reconstructions of *Archaeopteris* and further consideration of its phylogenetic position. *Amer. J. Bot.* **49**(4): 373–382.

BENNEK, CH. 1958. Die morphologische Beurteilung der Staub- und Blumen-blätter der Rhamnaceen. *Bot. Jahrb.* **77**: 423–457.

BENSON, F. M. 1904. *Telangium Scottii*, a new species of *Telangium* (*Calymmatotheca*) showing structure. *Ann. Bot.* **18**: 161–177.

BONNER, J., and J. A. D. ZEEVAART. 1962. Ribonucleic acid synthesis in the bud, an essential component of floral induction in *Xanthium*. *Plant Physiol.* **37**: 43–49.

BOS, L. 1957. Heksenbezemverschijnselen: een pathologisch-morfologisch onderzoek (Witches' Broom phenomena: A patho-morphological study). *Meded. Landbouwhogesch. Wageningen* **57**(1): 1–79.

BREMEKAMP, C. E. B. 1956. The concepts on which a morphology of the vascular plants should be based. *Acta Bot. Neerl.* **5**: 122–134.

———. 1962. The various aspects of biology: Essays by a botanist on the classification and main contents of the principal branches of botany. *Verhandel. Kon. Ned. Akad. Wetensch. Amsterdam Afd. Natuurk.* (Ser. 2) **54**(2): 1–199.

BUVAT, R. 1952. Structure, évolution et fonctionnement du méristème apical de quelques Dicotylédones. *Ann. Sci. Nat. Bot.* (11e Sér.) **13**: 199–300.

———. 1955. Le méristème apical de la tige. *Ann. Biol.* **31**: 595–656.

CAMP, W. H., and M. M. HUBBARD. 1963a. Vascular supply and structure of the ovule and aril in peony and of the aril in nutmeg. *Amer. J. Bot.* **50**: 174–178.

———. 1963b. On the origins of the ovule and cupule in lyginopterid Pteridosperms. *Amer. J. Bot.* **50**: 235–243.

CANRIGHT, J. E. 1952. The comparative morphology and relationships of the Magnoliaceae. I. Trends of specialization in the stamens. *Amer. J. Bot.* **39**: 484–497.

———. 1960. *Ibid.* III. Carpels. *Amer. J. Bot.* **47**: 145–155.

ČELAKOWSKY, L. 1874. Über die morphologische Bedeutung der Samenknospen. *Flora* **57**: 113–251.

———. 1875. Zur Discussion über das Eichen. *Bot. Ztg.* **33**: 193–201, 217–233.

———. 1876. Vergleichende Darstellung der Placenten in den Fruchtknoten der Phanerogamen. *Abh. böhm. Ges. Wiss.* (VI) **8**: 1–74.

———. 1899. Epilog zu meiner Schrift über die Plazenten der Angiospermen, nebst eine Theorie des antithetischen Generationswechsels der Pflanzen. *Sitzber. königl. böhm. Ges. Wiss., Math.-Naturw. Kl.* (*Prague*).

CHADEFAUD, M. 1946. L'origine et l'évolution de l'ovule des Phanérogames. *Rev. Sci.* **84**: 502–509.

——— and L. EMBERGER. 1960. *Traité de Botanique.* Vol. II (L. EMBERGER), *Les végétaux vasculaires.* Fasc. I. Paris.

CHEADLE, V. I. 1944. Specialization of vessels within the xylem of each organ in the Monocotyledoneae. *Amer. J. Bot.* **31**: 81–92.

———. 1953. Independent origin of vessels in the Monocotyledons and Dicotyledons. *Phytomorphology* **3**: 23–44.

———. 1955. The taxonomic use of vessels in the metaxylem of Gramineae, Cyperaceae, Juncaceae and Restionaceae. *J. Arnold Arb.* **36**: 141–157.

CHURCH, A. H. 1919. Thalassiophyta and the subaerial transmigration. *Oxford Bot. Mem. No. 3.*

CONSTANCE, L. 1955. The Systematics of the Angiosperms. In *A Century of*

Progress in the Natural Sciences, 1853–1953. Pp. 405–483. California Academy of Sciences, San Francisco.

CORNER, E. J. H. 1946. Centrifugal stamens. *J. Arnold Arb.* **27**: 423–437.

——. 1949a. The annonaceous seed and its four integuments. *New Phytologist* **48**: 332–364.

——. 1949b. The Durian theory or the origin of the modern tree. *Ann. Bot.* (N. S.) **13**: 367–414.

——. 1953. The Durian theory extended. I. *Phytomorphology* **3**: 465–476.

——. 1954a. *Ibid.* II. The arillate fruit and the compound leaf. *Phytomorphology* **4**: 152–165.

——. 1954b. *Ibid.* III. Pachycauly and megaspermy. *Phytomorphology* **4**: 263–274.

CROIZAT, L. 1947. *Trochodendron, Tetracentron* and their meaning in phylogeny. *Bull. Torrey Bot. Cl.* **70**: 60–76.

——. 1961. *Principia Botanica* (2 vols.).

DANSER, B. H. 1950. A theory of systematics. *Bibl. Biotheor.* **6**: 117–180.

DELEVORYAS, T. 1963. Investigations of North American Cycadeoids: Cones of *Cycadeoidea*. *Amer. J. Bot.* **50**: 45–52.

EAMES, A. J. 1931. The vascular anatomy of the flower with refutation of the theory of carpel polymorphism. *Amer. J. Bot.* **18**: 147–188.

——. 1951. Again: 'The New Morphology'. *New Phytologist* **50**: 17–35.

——. 1952. Relationships of the Ephedrales. *Phytomorphology* **2**: 79–100.

——. 1961. *Morphology of the Angiosperms*. New York, Toronto, London.

ECKARDT, TH. 1937. Untersuchungen über Morphologie, Entwicklungsgeschichte und systematische Bedeutung des pseudomonomeren Gynoeciums. *Nova Acta Leopoldina* (N. F. 5) **26**: 1–112.

——. 1957. Vergleichende Studien über die morphologischen Beziehungen zwischen Fruchtblatt, Samenanlage und Blütenachse bei einigen Angiospermen, zugleich als kritische Beleuchtung der 'New Morphology'. *Neue Hefte Morph.* **3**(2): 1–91.

EMBERGER, L. 1944. *Les Plantes fossiles dans leurs rapports avec les végétaux vivants. Éléments de paléobotanique et de morphologie comparée.* Paris.

——. 1949. Les Préphanérogames. *Ann. Sci. Nat. Bot.* (11e Sér.) **10**: 131–144.

——. 1952. Encore sur les Préphanérogames avec remarques générales sur la systématique. *Rec. Trav. Lab. Bot. Univ. Montp. Sér. Bot.* **6**: 11–14.

ERDTMAN, G. 1952. *Pollen Morphology and Plant Taxonomy: Angiosperms.* Stockholm.

——. 1960a. Pollen walls and Angiosperm phylogeny. *Bot. Notiser* **113**: 41–45.

——. 1960b. On three new genera from the Lower Headon Beds, Berkshire. *Bot. Notiser* **113**: 45–48.

FAGERLIND, F. 1944. Die Samenbildung und die Zytologie bei agamospermischen und sexuellen Arten von *Elatostema* und einigen nahestehenden Gattungen nebst Beleuchtung einiger damit zusammenhängender Probleme. *K. svenska Vetensk. Akad. Handl.* (Ser. 3) **21**: 1–30.

——. 1946. Strobilus und Blüte von *Gnetum* und die Möglichkeit, aus ihrer Struktur den Blütenbau der Angiospermen zu deuten. *Arkiv Bot.* **33a**(8): 1–57.

————. 1958. Is the gynoecium of the Angiosperms built up in accordance with the phyllosporous or the stachyosporous scheme? *Sv. Bot. Tidskr.* **52**(4): 421–425.

FLORIN, R. 1948. On the morphology and relationships of the Taxaceae. *Bot. Gaz.* **110**: 31–39.

————. 1951. Evolution in Cordaites and Conifers. *Acta Hort. Bergiani* **15**(11): 285–388.

————. 1954. The Female Reproductive Organs of Conifers and Taxads. *Biol. Rev.* (Cambr. Phil. Soc.) **29**: 367–389.

————. 1955. The Systematics of the Gymnosperms. In *A Century of Progress in the Natural Sciences,* 1853–1953. Pp. 323–403. California Academy of Science, San Francisco.

FRITSCH, F. E. 1945. Studies in the comparative morphology of the Algae. IV. Algae and archegoniate plants. *Ann. Bot.* (N. S.) **9**: 1–29.

GAUSSEN, H. 1943–55. Les Gymnospermes actuelles et fossiles. *Trav. Lab. Forest.* (*Toulouse*) I–X.

GERASSIMOVA-NAVASHINA, H. 1961. Fertilisation and events leading up to fertilisation and their bearing on the origin of the Angiosperms. *Phytomorphology* **11**: 139–146.

GIFFORD, E. M., Jr., and H. B. TEPPER. 1962a. Histochemical and autoradiographic studies of floral induction in *Chenopodium album. Amer. J. Bot.* **49**: 706–714.

————. 1962b. Ontogenetic and histochemical changes in the vegetative shoot tip of *Chenopodium album. Amer. J. Bot.* **49**: 902–911.

GOTHAN, W., and H. WEYLAND. 1954. *Lehrbuch der Paläobotanik.* Berlin.

GRANT, V. 1950. The protection of ovules in Flowering Plants. *Evolution* **4**: 179–201.

GRÉGOIRE, V. 1935. Sporophylles et organes floraux, tige et axe floral. *Rec. Trav. Bot. Néerl.* **32**: 453–466.

————. 1938. La morphogénèse et l'autonomie morphologique de l'appareil floral. I. Le carpelle. *La Cellule* **47**: 285–452.

HAAN, H. R. M. DE. 1920. Contribution to the knowledge of the morphological value and the phylogeny of the ovule and its integuments. *Rec. Trav. Bot. Néerl.* **17**: 219–324.

HAGEMANN, W. 1963. Weitere Untersuchungen zur Organisation des Sprossscheitelmeristems der Vegetationspunkt traubiger Floreszenzen. *Bot. Jahrb.* **82**: 273–315.

HAGERUP, O. 1934. Zur Abstammung einiger Angiospermen durch Gnetales und Coniferae. *Kgl. Dansk Vidensk. Selsk. Biol. Medd.* **11**: 1–83.

————. 1936. *Ibid.* II. Centrospermae. *Kgl. Dansk. Vidensk. Selsk. Biol. Medd.* **13**: 1–60.

————. 1938. On the origin of some Angiosperms through the Gnetales and the Coniferae. III. The gynaecium of *Salix cinerea. Kgl. Dansk. Vidensk. Selsk. Biol. Medd.* **14**: 1–34.

HAMANN, H. 1961. Merkmalsbestand und Verwandschaftsbeziehungen der Farinosae. *Willdenowia* **2**: 639–768.

————. 1962. Beitrag zur Embryologie der Centrolepidaceae mit Bemerkungen über den Bau der Blüten und Blütenstande und die systematische Stellung der Familie. *Ber. Deut. Bot. Ges.* **85**(5): 153–171.

HAMMEN, L. VAN DER. 1948. Traces of ancient dichotomies in Angiosperms. *Blumea* **6**: 290–301.

HARRIS, T. M. 1960. The origin of Angiosperms. *The Advancement of Science* **67**: 1–7.

——. 1961. The fossil Cycads. *Palaeontology (London)* **4**: 313–323.

HESLOP-HARRISON, J. 1952. A reconsideration of plant teratology. *Phyton (Ann. Rei Bot.)* **4**: 19–34.

——. 1958. The unisexual flower: A reply to criticism. *Phytomorphology* **8**(1–2): 177–184.

——. 1959. Growth substances and flower morphogenesis. *J. Linn. Soc. Bot. (London)* **56**: 269–281.

HILLMAN, W. S. 1962. *The Physiology of Flowering.* New York.

HJELMQVIST, H. 1948. Studies on the floral morphology and phylogeny of the Amentiferae. *Bot. Notiser Suppl.,* Vol. 2, page 1.

HOLTTUM, R. E. 1948. The spikelet in Cyperaceae. *Bot. Rev.* **14**: 525–541.

HUGHES, N. F. 1961a. Further interpretation of *Eucommiidites* Erdtman. *Palaeontology. (London)* **4**: 292–299.

——. 1961b. Geology. *Sci. Progr.* **49**: 84–102.

—— and R. A. COUPER. 1958. Palynology of the Brora coal of the Scottish Middle Jurassic. *Nature* **181**: 1482–1483.

HUTCHINSON, J. 1959. *The Families of Flowering Plants* (2nd. ed.). Vols. I and II. London.

JANCHEN, E. 1950. Die Herkunft der Angiospermen-Blüte und die systematische Stellung der Apetalen. *Österr. Bot. Z.* **97**: 129–167.

JEFFREY, C. 1962. The origin and differentiation of the archegoniate land plants. *Bot. Notiser* **115**: 446.

KARSTEN, G. 1918. Zur Phylogenie der Angiospermen. *Z. Bot.* **10**: 369–388.

LAM, H. J. 1935. Phylogeny of Single Features. *Gard. Bull. Str. Settl.* **9**: 98–112.

——. 1948. Classification and The New Morphology. *Acta Biotheor.* **8**: 107–154.

——. 1950. Stachyospory and phyllospory as factors in the natural system of the Cormophyta. *Sv. Bot. Tidsskr.* **44**: 517–534.

——. 1954. Again the New Morphology, elucidated by the most likely phylogeny of the female coniferous cone. *Sv. Bot. Tidsskr.* **48**: 347–360.

——. 1959a. Some fundamental considerations on the 'New Morphology'. *Trans. Bot. Soc. Edinb.* **38**: 100–134.

——. 1959b. Taxonomy: General Principles and Angiosperms. In W. B. TURRILL (ed.), *Vistas in Botany.* Pp. 3–75. London.

——. 1961a. Wezen en herkomst der Angiospermen *Versl. Gew. Verg. Afd. Natuurk., Kon. Akad. Wetensch. Amsterdam* **70**(8): 11–114.

——. 1961b. Reflections on Angiosperm phylogeny: facts and theories. *Proc. Kon. Akad. Wetensch. Amsterdam* (C) **64**: 251–276.

——. 1962. *Tradenda. Mijmeringen bij een afscheid.* Farewell Lecture, University of Leyden. Pp. 1–39. Leyden.

LEPPIK, E. E. 1955. Some viewpoints on the origin and evolution of Flowering Plants. *Acta Biotheor.* **9**: 45–56.

——. 1957. Evolutionary relationship between entomophilous plants and anthophilous insects. *Evolution* **11**: 466–481.

——. 1960. Early evolution of flower types. *Lloydia* **23**: 72–92.

LEROY, J. F. 1954. Étude sur les Juglandaceae. (Thèse, Fac. Sci. Univ. Paris.) *Mém. Mus. Nat. Hist. Nat.* **6**: 1–246.

LONG, A. G. 1960. On the structure of '*Samaropsis scotica*' Calder (emended) and 'Eurystoma angulare' gen. et sp. nov., petrified seeds from the calciferous sandstone series of Berwickshire. *Trans. Roy. Soc. Edinb.* **64**: 261–280.

McLean Thompson, J. 1934. Studies in advancing sterility. VII. The state of flowering known as angiospermy. *Publ. Hartley Bot. Lab.* (*Liverpool*) **12**: 1–48.

———. 1937. On the place of ontogeny in floral inquiry. *Publ. Hartley Bot. Lab.* (*Liverpool*) **17**: 3–20.

Maekawa, F. 1960. A new attempt in phylogenetic classification of plant kingdom. *J. Fac. Sci. Univ. Tokyo, Sect. III.* (*Bot.*) **7**: 543–569.

Maheshwari, P. 1950. *An Introduction to the Embryology of Angiosperms.* New York.

———. 1960. Evolution of the Ovule (Seventh Seward Memorial Lecture, Birbal Sahni Institute of Palaeobotany). Lucknow.

——— and V. Vasil. 1961a. The stomata of *Gnetum. Ann. Bot.* (N. S.) **25**: 313–319.

——— and ———. 1961b. *Gnetum. Bot. Monograph* (No. 1). New Delhi.

Majumdar, G. P. 1956. Carpel morphology. *J. Asiat. Soc. Sci.* **22**: 45–54.

Manning, W. E. 1938. The morphology of the flowers of the Juglandaceae. I. The inflorescence. *Amer. J. Bot.* **25**: 407–419.

———. 1940. *Ibid.* II. The pistillate flowers and fruit. *Amer. J. Bot.* **27**: 839–852.

———. 1948. *Ibid.* III. The staminate flowers. *Amer. J. Bot.* **35**: 606–621.

Meeuse, A. D. J. 1961a. The Pentoxylales and the origin of the Monocotyledons. *Proc. Kon. Ned. Akad. Wetensch. Amsterdam* (C) **64**: 543–559.

———. 1961b. Marsileales and Salviniales—'Living fossils'? *Acta Bot. Néerl.* **10**: 257–260.

———. 1962. The multiple origin of the Angiosperms. *Advancing Front. Plant Sci.* **1**: 105–127.

———. 1963a. The so-called 'megasporophyll' of *Cycas:* A morphological misconception. Its bearing on the phylogeny and the classification of the Cycadophyta. *Acta Bot. Néerl.* **12**: 119–128.

———. 1963b. From ovule to ovary: A contribution to the phylogeny of the megasporangium. *Acta Biotheor.* **16**: 127–182.

———. 1963c. Stachyospory, phyllospory and morphogenesis. *Advancing Front. Plant Sci.* **7**: 115–156.

———. 1964a. Some phylogenetic considerations of the process of double fertilisation. *Phytomorphology* **13**: 237–244.

———. 1964b. The bitegmic spermatophytic ovule and the cupule: A reconsideration of the so-called pseudo-monomerous ovary. *Acta Bot. Néerl.* **13**: 97–112.

———. 1965. Angiosperms, past and present. Phylogenetic botany and interpretative morphology of the Flowering Plants. *Advancing Front. Plant Sci.* **11**: 1–228. New Delhi.

——— and J. Houthuesen. 1964. The structure of the pistil of *Engelhardia spicata* (Juglandac.) and its phylogenetic significance. *Acta Bot. Néerl.* **13**: 352–366.

Melville, R. 1960. A new theory of the Angiosperm flower. *Nature* **188** (No. 4744): 14–18.

———. 1962. A new theory of the Angiosperm flower. I. The gynoecium. *Kew Bull.* **16**(1): 1–50.

———. 1963. *Ibid.* II. The androecium. *Kew Bull.* **17**(1): 1–63.

MERKER, H. 1961. Entwurf zur Lebenskreis-Rekonstruktion der Psilophytales nebst phylogenetischem Ausblick. *Bot. Notiser* **114**: 88–102.

METCALFE, C. R., and L. CHALK. 1950. *Anatomy of the Dicotyledons.* Oxford.

MOELIONO, B. M. 1959. A preliminary note on the placenta of *Stellaria media* (L.) Vill. and *Stellaria graminea* L. A possible axial origin of ovula? *Acta Bot. Neerl.* **8**: 292–303.

NEUMAYER, H. 1924. Die Geschichte der Blüte. *Abh. Zool. Bot. Ges. Wien* **14**: 1–112.

NOZERAN, R. 1955. Contributions à l'étude de quelques structures florales. (Essai de morphologie florale comparée.) *Ann. Sci. Nat. Bot.* (11ᵉ Sér.) **16**: 1–224.

OZENDA, P. 1948. Recherches sur les Dicotylédones apocarpique: Contribution à l'étude des Angiospermes dites primitives. (Thèse, Fac. Sci. Univ. Paris.)

PANKOW, H. 1959. Histogenetische Untersuchungen an der Plazenta der Primulaceen. *Ber. Deut. Bot. Ges.* **72**: 111–122.

———. 1962. Histogenetische Studien an den Blüten einiger Phanerogamen. *Bot. Stud.* **13**: 1–106.

PARKIN, J. 1957. The unisexual flower again: A criticism. *Phytomorphology* **7**: 7–9.

PAYER, J. B. 1857. *Traité d'organogénie comparée de la fleur.* Paris.

PEARSON, H. H. W. 1929. *The Gnetales.* Cambridge.

PERIASAMY, K., and B. G. L. SWAMY. 1956. The conduplicate carpel of *Cananga odorata. J. Arnold Arb.* **37**: 366–372.

PFLUG, H. D. 1953. Zur Entstehung und Entwicklung des angiospermoiden Pollens in der Erdgeschichte. *Palaeontographica* (Abt. B.) **95**: 60–171.

PHILIPSON, W. R., and E. E. BALFOUR. 1963. Vascular patterns in dicotyledons. *Bot. Rev.* **29**: 382–404.

PIJL, L. VAN DER. 1955. Sarcotesta, aril, pulpa and the evolution of the angiosperm fruit. *Proc. Kon. Ned. Akad. Wetensch. Amsterdam* (C) **58**(2): 154–161; **58**(3): 307–312.

———. 1960. Ecological aspects of flower evolution. I. Phyletic evolution. *Evolution* **14**: 403–416.

———. 1961. *Ibid.* II. Zoophilous flower classes. *Evolution* **15**: 44–59.

PLANTEFOL, L. 1946–1947. Fondements d'une théorie phyllotaxique nouvelle. *Ann. Sci. Nat. Bot.* (11ᵉ Sér.) **7**: 153–229 (1946); **8**: 1–71 (1947).

———. 1948. L'ontogénie de la fleur: Fondements d'une théorie florale nouvelle. *Ann. Sci. Nat. Bot.* (11ᵉ Sér.) **9**: 35–186.

PLUMSTEAD, E. P. 1952. Description of two new genera and six species of fructifications borne on *Glossopteris* leaves from South Africa. *Trans. Geol. Soc. S. Afr.* **55**: 281–328.

———. 1956a. Bisexual fructifications borne on *Glossopteris* leaves from South Africa. *Palaeontographica* (B) **100**: 1–25.

———. 1956b. On *Ottokaria*, the fructifications of *Gangamopteris. Trans. Geol. Soc. S. Afr.* **59**: 211–236.

POTONIÉ, H. 1912. *Grundlinien der Pflanzenmorphologie im Lichte der Paläontologie.* Jena.

POTONIÉ, R. 1959. The New Morphology (Fifth Seward Memorial Lecture, Birbal Sahni Institute of Palaeobotany, 1957). Lucknow.

PULLE, A. A. 1952. *Compendium van de Terminologie, Nomenclatuur en Systematiek der Zaadplanten* (3rd ed.). Utrecht.

PURI, V. 1951. The role of floral anatomy in the solution of morphological problems. *Bot. Rev.* **17**: 471–553.

———. 1960. On the methods of studying floral morphology. *Proc. Nat. Inst. Sci. India* **26**(B), (Suppl.): 97–108.

ROTH, I. 1957. Die Histogenese der Integumente von *Capsella bursa-pastoris* und ihre morphologische Bedeutung. *Flora* **145**: 212–235.

———. 1959. Histogenese und morphologische Deutung der Plazenta von *Primula*. *Flora* **148**: 129–152.

ROTHMALER, W. 1951–52. Die Gymnospermen und der Ursprung der Angiospermen. *Wiss. Z. Mart. Luther Univ.* **1**(4): 1–12.

SARGEANT, E. 1908. A reconstruction of a race of primitive Angiosperms. *Ann. Bot.* **22**: 121–186.

SATTLER, R. 1962. Zur frühen Infloreszenz- und Blütenentwicklung der Primulales *sensu lato* mit besonderer Berücksichtigung der stamen-petalum Entwicklung. *Bot. Jahrb.* **81**(4): 358–396.

SAUNDERS, E. R. 1937–39. *Floral Morphology: A new outlook with special reference to the interpretation of the gynoecium* (2 vols.). Cambridge.

SCOTT, R. A., E. S. BARGHOORN and E. B. LEOPOLD. 1960. How old are the Angiosperms? *Amer. J. Sci.* **258a**: 284–299.

SIMPSON, G. G. 1961. *Principles of Animal Taxonomy.* New York.

SPORNE, K. R. 1956. The phylogenetic classification of the Angiosperms. *Biol. Rev.* **31**: 1–29.

———. 1958. Some aspects of floral vascular systems. *Proc. Linn. Soc.* (*London*). 169th Session (1956–1957). Pts. 1 and 2, pp. 75–84.

———. 1959. The phylogenetic classification of plants. *Amer. J. Bot.* **46**: 385–394.

STEBBINS, G. L. 1951. Natural selection and the differentiation of Angiosperm families. *Evolution* **5**(4): 299–324.

SWAMY, B. G. L. 1950. *Sarcandra*, a vesselless genus of the Chloranthaceae. *J. Arnold Arb.* **31**: 117–129.

———. 1953. The morphology and relationships of the Chloranthaceae. *J. Arnold Arb.* **34**: 375–408.

———. 1963. The origin of cotyledon and epicotyl in *Ottelia alismoides*. *Beitr. Biol. Pflanz.* **39**(1): 1–16.

TAKHTAJAN, A. L. 1959a. *Essays on the Evolutionary Morphology of Plants.* (Transl. by O. H. Gankin.) American Institute of Biological Sciences.

———. 1959b. *Die Evolution der Angiospermen.* Jena.

TEPFER, S. S. 1953. Floral anatomy and ontogeny in *Aguilegia formosa* var. *truncata* and *Ranunculus repens*. *Univ. Calif. Publ. Bot.* **25**(7): 513–648.

THOMAS, H. HAMSHAW. 1931. The early evolution of the Angiosperms. *Ann. Bot.* **45**: 647.

———. 1936. Palaeobotany and the origin of the Angiosperms. *Bot. Rev.* **2**: 397–418.

TIEGHEM, PH. VAN. 1875. Recherches sur la structure du pistil et sur l'anatomie comparée de la fleur. *Mém. Acad. Sci. Inst. Imp. de France* **21**: 1–261.

TOWNROW, J. A. 1962. On *Pteruchus*, a microsporophyll of the Corystosper-maceae. *Bull. Brit. Mus.* (*Nat Hist.*), *Geol.* **6**(2): 287–320.

TROLL, W. 1935, 1937–39. *Vergleichende Morphologie der höheren Pflanzen.* Berlin.

VISHNU-MITTRE. 1953. A male flower of the Pentoxyleae, with remarks on the structure of the female cones of the group. *Palaeobotanist* **2**: 75–84.

———. 1957. Studies on the fossil flora of Nipania (Rajmahal Series), India —Pentoxyleae. *Palaeobotanist* **6**: 31–46.

WALTON, J. 1953. *An Introduction to the Study of Fossil Plants* (2nd ed.). London.

WARDLAW, C. W. 1952a. Methods in plant morphogenesis. *J. Linn. Soc. London (Bot.)* **56**: 154–160.

———. 1952b. *Phylogeny and Morphogenesis.* London.

———. 1955. *Embryogenesis in Plants.* London.

———. 1960. The inception of shoot organisation. *Phytomorphology* **10**: 107–110.

———. 1961. Growth and Development of the Inflorescence and Flower. In M. X. ZARROW (ed.), *Growth in Living Systems* (International Symposium on Growth, Purdue University, 1960). New York.

WETTSTEIN, R. VON. 1935. *Handbuch der Systematischen Botanik* (4. Aufl.). Leipzig and Vienna.

WIEBES, J. T. 1963: Taxonomy and host preferences of Indo-Australian fig wasps of the genus *Ceratosolen* (Agaonidae). *Tijdschr. Entomol.* **106**: 1–112. (Doctoral thesis, Leiden, 1963.)

WILSON, C. L. 1937. The phylogeny of the stamen. *Amer. J. Bot.* **24**: 686–699.

———. 1942. The telome theory and the origin of the stamen. *Amer. J. Bot.* **29**: 759–764.

WOLFF, C. F. 1759. Theoria generationis. In Ostwald, *Klassik der exakten Wissenschaften.* Pp. 84, 85. Leipzig.

YOSHIDA, O. 1959. Embryologische Studien über die Ordnung Piperales. III. Embryologie von *Sarcandra glabra* Nakai. *J. Coll. Arts Sci. Chiba Univ.* **3**: 55–60.

ZIMMERMANN, W. 1957. Phylogenie der Blüte. *Phyton* **7**(1–3): 162–182.

———. 1959. *Die Phylogenie der Pflanzen* (2. Aufl.). Stuttgart.

———. 1965. *Die Telomtheorie. Fortschritte der Evolutionsforschung,* Bd. I. Stuttgart.

Author Index

Index of Scientific Names

221

Subject Index

Main discussion is centred in passages designated by italicised page numbers.